Drug Abuse in Adolescence

Denise De Micheli • André Luiz Monezi Andrade •
Eroy Aparecida da Silva • Maria Lucia Oliveira
de Souza Formigoni
Editors

Drug Abuse in Adolescence

Neurobiological, Cognitive, and Psychological Issues

Springer

Atheneu

Editors
Denise De Micheli
Universidade Federal de São Paulo
São Paulo
Brazil

Eroy Aparecida da Silva
Universidade Federal de São Paulo
São Paulo
Brazil

André Luiz Monezi Andrade
Universidade Anhembi Morumbi
São Paulo
Brazil

Maria Lucia Oliveira de Souza-Formigoni
Universidade Federal de São Paulo
São Paulo
Brazil

This work was originally published in Portuguese by Editora Atheneu LTDA under the title of Neurociências do abuso de drogas na adolescência—O que sabemos?

ISBN 978-3-319-17794-6   ISBN 978-3-319-17795-3 (eBook)
DOI 10.1007/978-3-319-17795-3

Library of Congress Control Number: 2015943312

Springer Cham Heidelberg New York Dordrecht London

Printed on acid-free paper

Springer International Publishing AG Switzerland is part of Springer Science+Business Media (www.springer.com)

To God, Whom I get to know more every day and Who increasingly guides my life. To Him, my love, my gratitude, talent, inspiration, breath of life!

To the one who is always by my side. My love, companion, friend, partner: Nicolau. My main supporter. Ours is a bond of love and partnership. We are better together!

To the reason why there is a sparkle in my eyes: Nicolas and Carla! My life is more colorful and wonderful every day thanks to you.

To my parents, Regina and Gianluigi. My heart is full of love and gratitude to you. Examples of self-giving, integrity and honor.

Denise De Micheli

To my parents, Irene and Sebastião, for their teachings on the art of living.

To Daniel, my beloved son, whose light lit the path.

Eroy Aparecida da Silva

To my parents, Luiz Otávio and Marta, who taught me the value of work and love.

To the teachers who provided the basis for my education, the pupil of whom I had the honor to be: Alfredo Lohr Junior e Andréia Schmidt.

André Luiz Monezi Andrade

To Gilberto, Daniela, and Carla, the loves of my life, for their affection, support, encouragement, and understanding.

To Fabio and Cristina, hoping they'll find the serenity needed to understand the changes that occur in adolescence and the wisdom to guide their children through this period of growth and transition.

To my nephews, Lucas and Bruno, to whom I wish a happy adolescence, with intelligent and healthy choices.

Maria Lucia Oliveira de Souza Formigoni

Rubem Alves' Garden

"Poetic experience does not mean to see wonderful things no one else sees. It is to see the most trivial, that which is right under one's nose, under a different light."

I believe that upon creating the world, God had one single word in His mind: garden! A garden is the image of beauty, harmony, love, and happiness. If I were to say one last word, one single word, it would certainly be "garden."

After a long wait, I was finally able to sow my garden. I had a long wait because gardens need land to exist. But I had no land. The only thing I had was a dream. I know that gardens exist in dreams well before they exist outside. A garden is like a dream that has become reality, a revelation of our internal, hidden truth, the naked soul shamelessly offering itself for the pleasure of others.... But dreams alone can do nothing: they are like birds without wings.... They are like songs, which are nothing until someone sings them; like seeds, within little packages, waiting for someone to free them and sow them in the ground. My dreams lived inside me. They were mine. But the land was not mine.

The land was next to my house, narrow, without any space, between walls. It was a wasteland, full of garbage, bushes, thorns, broken glass, rusty tins; a place inhabited by scary rats that would visit us once in a while. When my dream was too tight inside my heart, wishing to come alive, I would place the ladder against the wall to get a glimpse of the lot. I stayed there, imagining my garden. I kept sowing my seeds and watching them bud … merely in my mind.

At times, I did not believe that my dream could come true. Eventually, I started looking to move to another house because it was clear that other people had other plans for that land in which my dreams lived. Were the dreams of others to come true, I would become like a bird in a cage, pressed against two walls, sentenced to a life of misery.

However, one day, the unexpected happened. The land became mine. My dream made love to the land, and the garden was born. Therefore, I can say that each and every garden starts out as a love story. Before any tree is sown or lake fills, they must first be born inside one's soul. Whoever does not sow a plant inside will not sow gardens outside and, thus, will not be able to walk in them.

Rubem Alves

Introductory Quotation

We sowed a garden in our dreams over a very long time. And some time ago, that dream came true, and our garden became a reality.

This book represents one more seed for our garden—that garden we sowed in which to watch our ideas become fertile, bud, sprout … and then grow and take shape. We sowed a garden in which ideas could blossom through us … but not necessarily in us.

We hope that you will benefit from reading this book and that seeds of ideas might simultaneously be born in your mind and heart because, if they remain as thoughts only, they will never be more than mere good ideas!

We take Rubem Alves' words as ours:

> I'm sowing the seeds of my highest hope. I'm not looking for disciples to communicate my knowledge to them. Knowledge is freely available to anyone who wants it. I'm looking for disciples to sow my best hopes in them.

Have a pleasant reading.

Denise De Micheli
André Luiz Monezi Andrade
Eroy Aparecida da Silva
Maria Lucia Oliveira de Souza-Formigoni

Preface

Hamar Nastasy Palhares Alves

Neuroscience of Drug Abuse in Adolescence: What Do We Know?

The fact that prefaces often begin by describing how difficult it is to write one should have been a warning. But prefaces are there, and their existence is a reminder that trying to escape from praising a book is much more difficult than writing one.

To write the preface to a book is a complex human behavior. Sedimented on variegated motivations, affinities, and skills, it's an honor and something that makes its author proud. But it is not, and cannot be, just that. By "complex," I mean, "unique," without any obvious, reproducible recipe.

Behavior, the mind, the human soul—pick what you like best—has a deep and intricate nature. Linearity is one of the fallacies of human behavior. In a linear world, man would be a drawing on a sheet of paper, easily enclosed by a circle around him. That is the world of yes and no. Not being constrained to a 2D world is reassuring; nevertheless, we often fall back into the "all or nothing" world. Let me explain. You must have already watched interview shows in which some celebrity is required to opine on a recent event, like a soccer championship or a massacre: "Do you believe that on-screen violence might have influenced those kids?" One interviewee answers "yes," whereas the other swears to the contrary. The discussion heats up, and the presenter announces an advertising break; you go grab a snack from the fridge while awaiting the next round.... The point is that both opinions are wrong. But they are also right. Behavior is complex and follows the laws of multicausality; for that reason, behavior unfolds within the vast gap that extends from "never" to "always."

As humans, we walk together along a long and preset path. As individuals, however, our paths are singular and unique.

A part of that path, one of the most beautiful and unforgettable ones is conventionally named "adolescence." The notion of adolescence is recent phenomenon, created by Western men. As a concept, adolescence is still in its cradle.

A period of synesthetic experiences, of "friends for life"; intense, hot, with the onset of strong, powerful passions; the discovery of the petals of sensuality, of the rapture of bodily sensations. Of the oh-so-welcome butterflies in the stomach. Of fights at schools, threats, rebelling against the ones we love most, the shaping of new social groups, cliques, gangs. To pull an all-nighter studying. Or at the nightclub. Of experiences fueled by the 24/7, non-stop flow of online social networks. And, like it or not, the most circumspect experiences: the time when the use of alcohol, tobacco, and other drugs starts in real life. Although not systematically nefarious, the lives of many will be deeply touched, if not actually interrupted, by that encounter. But the consequences can be (somewhat) different. This is what we shall see in the present journey.

The book you now have in your hands and are beginning to leaf through shyly, dear reader, is a well-marked path composed of dense, tempting, and viscerally well-crafted essays on the interface among neuroscience, adolescence, and drug use. You can start with any chapter, but if I may make a suggestion, read it as if you were listening to a concept album, like *The Dark Side of the Moon* or *Sgt. Pepper's*, because the sequence of the first essays makes the following ones richer. Notably, each voice in this book has its due time of discourse, which grants them density, reach, expressiveness, and reciprocally, bestowing a beneficial "human touch" to the book.

The word "interface" is apt. It's enticing, sexy, tempting. At the same time, it is a vow of humbleness. It represents an acknowledgment of the fact that the human experience demands several skills and fields of knowledge to achieve broad and deep-rooted understanding. It is a signal that investigators are exploring the living matter of change. To place oneself at an interface is to place oneself at the edge of an abyss: to assert that what we do not yet know is as important as that we already know. But it is precisely there that the wind of inspiration blows the strongest, where the landscape is precious and delightful. It is there that the truly unforgettable portraits of the soul are grasped.

Neuroscience is evocative of a primordially instigating and fascinating field of studies; it provides us tools to understand that which makes us common and viable, that which makes us unique, erring, and exciting. Several focal levels coexist in the understanding of behavior, attitudes, and the uniqueness of each individual; as will be demonstrated, however, it is not a matter of molecules dictating social phenomena. Rather, it is one and the other at the same time, in multiple concurring and inter-fertilizing planes of experience.

When I was a medical resident, one patient often came to me to ask: "Can a remedy change the way I experience the world?" I'd become stunned and give him the explanations my teachers had taught me. The fact is, I admit, that I was confused.

Yet, a remedy might help a person stand on his or her feet again—exactly as a drug might make him or her lose his or her mind. But neither a remedy nor a drug will define what the individual will do with his or her head or feet.

In the pages of this book, dear reader, you will be invited to explore a highly and exquisitely elaborated set of reactions and relationships that shape, expand, depress, and, unfortunately, usually put the odyssey of living in chains. The book describes a self-inflicted type of slavery that is paid with one's own freedom to choose. However, because we are humans, there will always be opportunity. And one such opportunity is that for getting our freedom back.

A *zen* text compares man to a vase delicately brought into warmth and light by the potter's hands. As the clay dries, the vase acquires it final shape, and little flexibility will then remain in it. Changes will only concern its polishing and painting, the use that will be made of it—they will not affect its intimate structure. This is a significant part of the (simple and rich) view neuroscience has relative to the shaping of personal development. The stages with the greatest potentiality and plasticity are childhood and adolescence, when continuous processes involving new knowledge and challenges condense, expand and chisel the personality, bestowing shine and suffering on it. However, because we are humans, some flexibility will always remain.

One rich metaphor used in Buddhist psychology compares the subconscious to a granary, where non-manifested attitudes wait in silence. All of the seeds wait dormant in the endless and multiform warehouse of the self. From the ground of actual and manifested behavior springs that which we sprinkle with water, and also that which we did not take care enough to pluck off. Pruning defines that which will grow in a more exuberant manner. Watering is necessary. Watering, pruning, and pulling away are verbs that serve to assert who we are, what we can and might be, and what we would like to change.

The time has arrived in this preface, which might have bored the readers more than was expected, to praise the magnificent effort to bring this book to the public, which is priceless as a function of the careful research that underpins each essay. In addition to having been sophisticatedly orchestrated by the chief maestros, Denise De Micheli, André Luiz Monezi Andrade, Eroy Aparecida da Silva, and Maria Lucia Oliveira de Souza-Formigoni, the essays, at the necessary and optimal level, make us feel that we are bestowed new, applicable and useful knowledge, without being taken to a lofty and piddling world of secret, unprofitable or inaccessible information. Warm bread freshly taken from the oven: this is what you have here, dear reader. Once the book has been read, we will have the pleasant feeling of having treaded a coherent, enriching, and luminous path. And it will make us come back to it in the search of inspiration for further journeys and coziness amidst the troubled waters of so many pieces of knowledge disconnected from the hyper-connectivity in which we are immersed.

I'd like to finish by relating an experience from my own adolescence. Several crossroads had multiplied, and the associated problems had become intertwined: a dear aunt, Martha, hurried to help me. That particular crisis was characterized by a singular factor of aggravation, which made it especially dramatic: it was about myself! I did not want to fit in any model or statistics. No adolescent ever wants so. Nothing that would stretch me or shrink me within already worn models. Recognizing these limits, my aunt was sensitive in bringing one of the most precious sections of the Bible to me, from the Book of Ecclesiastes:

> To every thing there is a season, and a time to every purpose under the heaven: a time to be born, and a time to die; a time to weep, and a time to laugh; a time to embrace, and a time to refrain from embracing; a time to love, and a time to hate; a time of war, and a time of peace.

That fragment copied by hand brings comfort, appeasement and focus to me to this day. It was tattooed in multiple neurons and minute crevices that make me feel and live as I am: sometimes, one loving word at the right time is all a youth might need.

Adolescence is the time of the ardent flame, and that which distinguishes us from adolescents is that the latter do not forget that now is the time to live.

For you, my dear companion, valued reader, it is the time to enjoy this precious journey that begins on the next pages. And, yes, to live! Enjoy!

Hamer Nastasy Palhares Alves
Psychiatrist

Preface

Sidarta Tollendal Gomes Ribeiro

Since the very beginnings of civilization, humankind has had an intimate relationship with drugs, which are used to heal the body and the mind from their many pains. However, to the best of our knowledge, there was not such tremendous abuse of drugs in the past as in the present. That problem has become increasingly more severe in parallel to the vertiginous growth of the world population, whereby the deepest human contacts become rarer and, consequently, the mechanisms of social regulation degrade. Uprooted, deritualized, and dislodged from their traditional contexts of use, drugs especially affect the least informed and most impulsive individuals, woefully ravaging the youngest ones. Adolescence is a period of major discoveries and adventures; however, it is also a time when existential instability, a lack of motivation to comply with adult norms, diffuse rebellion against authority and peers' ethical and esthetic pressures exert the greatest power. Several significant choices are made at that age, most of them as the result of trial and error. It is at that crossroads in life that so many adolescents enter into contact with drugs, and many of them get hurt.

Unfortunately, society offers very little help and support to youngsters. A culture of prohibition still rules over almost the entire continent, hindering the free circulation of ideas, blocking candid conversation, and promoting paranoia, at the expense of damage control. Education aiming at the responsible use of drugs is incipient, from alcohol to sugar, from marijuana to benzodiazepines; instead, the emphasis falls on the pure, simple, inefficacious, and excluding method of repression, which does not distinguish among classes of substances, ignores neuroscience, and does not acknowledge social contexts. For all of those reasons, our society strikingly fails in its attempts to reduce the suffering associated with drugs.

The antidote against such major disaster is known as high-quality information. This new book, *Neuroscience of Drug Abuse in Adolescence: What Do We Know?*, fills in significant gaps relative to the neurobiological, cognitive, and psychological aspects of drug abuse and addiction in adolescence. The book editors, Denise De Micheli, André Luiz Monezi Andrade, Eroy Aparecida da Silva, and Maria Lucia Oliveira de Souza-Formigoni, dissected and mounted a vast corpus of multidisciplinary literature to paint a broadly scoped picture of the mechanisms that underlie chemical dependency in adolescence. The book chapters, written by reputed experts, discuss legal and illegal drugs in their complex relationships with mental diseases and disorders. The book is organized along two axes focused on adolescence: "Psychobiological Transformations" and "Drugs and the Central Nervous System." Epidemiological, chronobiological, endocrinological, neurochemical, physiological, and neurological aspects in human beings, as well as in animal models of adolescence, are discussed. Disorders of mood, motivation, and cognition are particularly significant.

Although its main thrust is coherent, the book is not monolithic as to the social implications of the biological findings. Although some chapters advocate a more traditional view on the interpretation of associations and comorbidities, others make it quite clear that cause and correlation are different. As a whole, the book reflects a time of transition, in which science is increasingly rejecting prohibitionist bias to begin facing the problem posed by drug abuse without preconceptions and, in particular, to focus on the early use of drugs.

This is a serious problem—a true, complex, and multifaceted problem in both monstrous megalopolises and boondocks lost to the twenty-first century. We must confront it with our eyes wide open, armed with as much information as possible, as well as with the necessary component of utopia that allows us to hope that we will learn how to coexist with all types of drugs, as revealed in their manifold relations to users, their relatives, and clinicians. No drug is pure evil. Pure evil is the ignorant use of so much power.

Sidarta Tollendal Gomes Ribeiro
Neuroscientist, Full Professor and Director,
Brain Institute, Federal University of Rio Grande do Norte

Contents

Contributors

Monica Levy Andersen Departamento de Psicobiologia, Universidade Federal de São Paulo, São Paulo, Brazil

André Luiz Monezi Andrade Department of Psychology, Anhembi Morumbi University—*Laureate International Universities*®, São Paulo, Brazil

Marco Antonio Bessa Federal University of São Paulo/School of Medicine of São Paulo (UNIFESP/EPM), São Paulo, Brazil

Roseli Boerngen-Lacerda Department of Pharmacology, Federal University of Paraná (UFPR), Curitiba, Brazil

Helena Maria Calil Department of Psychobiology, UNIFESP/EPM, São Paulo, Brazil

Rosana Camarini Department of Pharmacology, Institute of Biomedical Sciences, USP, São Paulo, Brazil

Silvana Pochetto Cracasso Federal University of São Paulo (UNIFESP), São Paulo, Brazil

Monique Ernst Section on Development and Affective Neuroscience, National Institute of Mental Health Intramural Research Program, Bethesda, MD, USA

Cassandra D. Gipson Department of Neuroscience, Medical University of South Carolina, Charleston, USA

Óscar Filipe Gonçalves Laboratory of Neuropsychophysiology, School of Psychology, University of Minho, Braga, Portugal; Spaulding Neuromodulation Center, Department of Physical Medicine & Rehabilitation, Spaulding Rehabilitation Hospital and Massachusetts General Hospital, Harvard Medical School, Charlestown, MA, USA; Applied Psychology, Bouvé College of Health Sciences, Northeastern University, Boston, MA, USA

Francine de Oliveira Goeldner Curitiba, Brazil

Luís Pereira Justo Associação Fundo de Incentivo à Pesquisa, São Paulo, Brazil

Peter W. Kalivas Department of Neuroscience, Medical University of South Carolina, Charleston, USA

Alfredo Löhr Center of Health Sciences, Pontifical Catholic University of Paraná (Pontifícia Universidade Católica do Paraná—PUCPR), Curitiba, Brazil

Valentina Lorenzetti Melbourne Neuropsychiatry Centre, Department of Psychiatry, The University of Melbourne, Parkville, VIC, Australia; Monash Clinical and Imaging Neuroscience, School of Psychological Sciences & Monash Biomedical Imaging, Monash University, Clayton, VIC Australia

Ana Raquel Mesquita Laboratory of Neuropsychophysiology, School of Psychology, University of Minho, Braga, Portugal

Denise De Micheli Department of Psychobiology, Universidade Federal de São Paulo—UNIFESP, São Paulo, Brazil

Ricardo M. Pautassi National Council of Scientific and Technological Research, Buenos Aires, Argentina; Department of Psychology, National University of Córdoba, Córdoba, Argentina

Fernanda Peralta School of Medicine, University of São Paulo (USP), São Paulo, Brazil

Gabriel Natan Pires Departamento de Psicobiologia, Universidade Federal de São Paulo, São Paulo, Brazil

Rista C. Plate Department of Psychology, University of Wisconsin-Madison, Bethesda, MD, USA

Elton Rezende Federal University of Minas Gerais, Belo Horizonte, Brazil; Federal University of São Paulo (UNIFESP) and University of São Paulo (USP), São Paulo, Brazil

Rafaela Larsen Ribeiro Department of Psychobiology, Federal University of São Paulo (UNIFESP), São Paulo, Brazil

Jessica M. Richards Department of Psychiatry and Human Behavior, Johns Hopkins University School of Medicine, Baltimore, MD, USA

Adriana Sampaio Laboratory of Neuropsychophysiology, School of Psychology, University of Minho, Braga, Portugal

Eroy Aparecida da Silva Associação Fundo de Incentivo à Pesquisa (AFIP), São Paulo, Brazil

Maria Lucia Oliveira de Souza-Formigoni Department of Psychobiology, Escola Paulista de Medicina, Universidade Federal de São Paulo—UNIFESP, São Paulo, Brazil

Michael Takagi Clinical Sciences, Murdoch Childrens Research Institute, Melbourne, Australia

Sergio Tufik Departamento de Psicobiologia, Universidade Federal de São Paulo, São Paulo, Brazil

Maria Sylvia de Souza Vitalle UNIFESP/EPM, São Paulo, Brazil

Shirley Yamaguchi Institute of Psychiatry, School of Medicine, University of São Paulo (USP), São Paulo, Brazil

George Youssef School of Psychology, Deakin University, Australia; Monash Clinical and Imaging Neuroscience, School of Psychological Sciences and Monash Biomedical Imaging, Monash University, Australia

About The Editors

Denise De Micheli BS in Psychology. PhD and postdoctoral fellowship in Science, Federal University of São Paulo (*Universidade Federal de São Paulo*—UNIFESP). Associate Professor and Chair of Medicine and Sociology of Drug Abuse, Department of Psychobiology, UNIFESP. Coordinator, Interdisciplinary Center for Studies in Neuroscience, Health and Education in Adolescence (*Centro Interdisciplinar de Estudos em Neurociência, Saúde e Educação na Adolescência*—Ciensea). Member of the Supervisory Board of the Research Incentive Fund Association (*Associação Fundo de Incentivo à Pesquisa*—AFIP). Member, Association for Medical Education and Research in Substance Abuse (Amersa). Collaborator, Institute for Translational Research in Adolescent Behavioral Health.

André Luiz Monezi Andrade Graduated in Psychology, Positivo University. Msc and PhD in Psychobiology, Federal University of São Paulo/School of Medicine of São Paulo (UNIFESP/EPM—*Escola Paulista de Medicina*). Researcher in Psychobiology of drug dependence professor at the Department of Psychology from Anhembi Morumbi University—*Laureate International Universities*®.

Eroy Aparecida da Silva BS in Psychology, São Paulo Organization of Education and Culture, Paulistana College. Specialist in Social and Occupational Psychology, Sedes Sapientiae Institute. Specialist in Family and Marriage Counseling, Pontifical Catholic University of São Paulo (*Pontifícia Universidade Católica de São Paulo*—PUC-SP). Doctor of Science, Department of Psychobiology, Federal University of São Paulo—UNIFESP. Full member, Brazilian Multidisciplinary Association of Studies on Drugs (*Associação Brasileira Multidisciplinar de Estudos sobre Drogas*—ABRAMD).

Maria Lucia Oliveira de Souza-Formigoni BS in Biological Sciences—biomedicine, Federal University of São Paulo/School of Medicine of São Paulo (UNIFESP/EPM). MS and PhD in Pharmacology, concentration in Psychopharmacology; Habilitation in Psychobiology, UNIFESP/EPM. Member, Research Society on Alcoholism, ISBRA (*International Society of Biomedical Research on Alcoholism*) and LASBRA (*Latin American Society of Biomedical Research on Alcoholism*) Assistant editor of the journal *Addiction*.

Part I
Adolescence: Psychobiological Transformations

Chapter 1
The Neurohormonal System in Adolescence

Maria Sylvia de Souza Vitalle

The Neurohumoral System in Adolescence

Puberty is a period that begins between childhood and adulthood characterized by intense body transformations; it is a part of the stage of life known as adolescence. Therefore, adolescence is a critical period of development signaled by change.

Several studies [1, 2] (including those using electrophysiological, imaging, pharmacological, and reactivity methods among others) have demonstrated that the brains of adolescents differ from both child and adult brains regarding their morphological and functional features as well as their structures, regions, circuits, and systems. In addition, adolescent brains differ with regard to their gray and white matter, their connectivity among structures, and their neurotransmission.

Adolescence, especially the peripubertal period, is a stage of remarkable development that includes significant brain changes such as synaptic pruning, the emergence of new fibers, and myelination. The different maturation rates of the brain areas related to emotional regulation and executive function might partially account for the remarkable increase in adolescents' engagement in high-risk situations and the search for new situations [3]. Morphophysiological changes occur that derive from the reactivation of the neurohormonal mechanisms in the hypothalamic-pituitary-gonadal axis, which are a part of a *continuum* that begins during intrauterine life and finishes when growth and development are complete. The neuroendocrine system regulates this process and determines its onset, the time required for maturation to occur, and the rate of change.

The age at which puberty is triggered is highly variable across individuals. Although a pattern of pubertal development exists, its actual progression is subjected to several influences, including genetic, ethnic, and racial factors; psychosocial and socioeconomic factors; climatic and geographic factors; the presence or aggravation of diseases; exercise; and the use of drugs or medication [4].

M. S. de S. Vitalle (✉)
UNIFESP/EPM, São Paulo, Brazil
e-mail: vitalle.dped@epm.br; sylviavitalle@gmail.com

D. De Micheli et al. (eds.), *Drug Abuse in Adolescence*,
DOI 10.1007/978-3-319-17795-3_1

The central nervous system, especially with regard to brain anatomy and function, undergoes substantial changes from late childhood to adolescence. These changes affect adolescent behavior, are genetically mediated, and are developed during the intrauterine period (the first period of major transformations when millions of synapses are formed) based on the exposures to which individuals are subjected throughout their lives.

One might assert that puberty is the stage of life during which the skills to solve complex problems in a mature manner emerge. The brain, subject to remodeling and learning over at least 10 years, undergoes a massive reorganization that might be considered the basis of the attitudes exhibited by adolescents. Along this process, brain structures and functions undergo the refinement needed for neurocognitive, affective, and social maturation. Such refinement occurs in a progressive manner and along a definite direction: from the posterior portion of the brain to its frontal portion over several years. The occipital lobe, located in the brain's posterior area, is responsible for the most basic functions such as visual perception and processing. The frontal portion is responsible for more complex functions, and it is the last portion to mature, accounting for the most elaborate cognitive and emotional functions such as planning, the production of mental representations of the external world, logical reasoning, and speech production. The cerebellum is related to motor control and learning, the maintenance of the body balance and posture, the perception of music and mathematics, and advanced social skills. The cerebellum exhibits high plasticity and is sensitive to external stimuli; its number of neurons and complex connections increases throughout adolescence [5].

Importantly, the amygdala, which is located in a still-developing prefrontal cortex, is where primary emotions such as fear and anger are activated. Therefore, this structure might account for the impulsiveness and behavioral maladjustments specific to adolescence. Neuronal mechanisms significantly participate in the elaboration of emotions while the adolescent brain undergoes significant changes [6]. Some neuronal evidence indicates an association between the brain's defense systems and the fear-stress-anxiety concept. The reactions or responses of orientation when facing (real or delusory) signals of danger, avoidance, or preparation with regard to confronting danger seem to be associated with anxiety. This process involves the participation of the cyngulate gyrus and the prefrontal cortex on one side as well as the contralateral median raphe nucleus, septum, and hippocampus, which are a part of the brain circuits that integrate emotional responses. Fearful stimuli that trigger active defense patterns restricted to specific situations but that are poorly elaborated induce other emotional states and seem to be associated with basic manifestations of fear. The dorsal periaqueductal gray matter is the major neural substrate for the integration of these brain circuits, and it participates in fear/anxiety responses [7–9].

We make choices all the time; the same is true of adolescents. However, the choices that adolescents make are not always the best response to a given situation because they have not yet processed the repertoire of cognitive-social-emotional experiences that enable them to make the best choices. An awareness regarding the irreversibility of the numerous choices one must make progressively develops over the course of adolescence.

Adolescence is a unique time to explore existential matters such as life and death [10]. In fact, death is one of the causes of fear that adolescents often mention [11–13]. The amygdala also participates in the "reward center" that also includes the hypothalamus, basal ganglia, and certain thalamic areas related to compulsive behavior.

The cortex metabolic rate is high throughout childhood, and it begins to decrease during late childhood to reach adult levels during the second decade of life. Simultaneously, the prefrontal cortex, which is related to the planning and regulation of emotional behavior, continues developing until approximately 20 years old. The anatomical changes that occur in the prefrontal cortex during physical maturation are most likely related to many of the behavioral changes observed throughout puberty [14, 15]. Linear volumetric increases in white matter occur between 4 and 22 years old as a function of the myelination that occurs throughout development, whereas gray matter undergoes more complex changes [16, 17]. Between 2 and 16 years old, cortical synaptic density and neuronal density decrease, and programmed cell death occurs, such that programmed synaptic reduction is associated with the improvement memory [18, 19].

The temporal lobe is related to memory, hearing, and the processing of visual and auditory information; it grows until reaching its maximum by 16–17 years old; greater neuronal communication during this period is favorable for reading activities.

The parietal lobe receives and processes spatial and sensory information from the entire body. Importantly, by 10–12 years old (i.e., the time when most puberty-related body changes occur), the parietal lobe attains its largest volume. Physical appearance plays a significant role in the development of adolescents' self-concepts. During this new process of refinement, an enormous effect on adolescent self-image might take place to promote the adjustment of sensory-motor "maps" of new body proportions.

The changes in the brain's reward system and the newly acquired ability for abstract thinking are the guiding impulses for adolescents to forego old childhood habits and develop interests in music, sports, religion, and philosophy. We might assert that the acquisition of new skills and the development of abstract reasoning lead adolescents to question rules and discover the social, economic, and cultural complexity of life. For that reason, they become impatient, forsake childish pleasures and the safety of their parents' home to look for satisfaction and pleasure in new activities, seek new friends, and expose themselves to risks they might not have even imagined before. The emergence of an explicit interest in sex is related to their hypothalamic-pituitary activity and the release of sexual hormones that increase during puberty.

Diseases such as mania, depression, obsessive-compulsive behavior, and schizophrenia are more common after puberty. That fact might indicate a relationship between these diseases and abnormalities concerning the architectural and functional changes of the brain throughout puberty [20]. A pattern of gray matter loss has been described in certain childhood psychoses that progress along the same direction as brain development during adolescence (i.e., from the posterior to anterior portions).

This pattern corresponds to the structural changes in gray matter characteristic of normal brain development, albeit at an abnormally high rate. Therefore, a pattern of disease reflects a lack of control in the regulation of gray matter maturation [15].

The major biological transitions that adolescents undergo likely make them more vulnerable to stress and the onset of psychopathologies. Stressful experiences during childhood and adolescence are associated with the development of later psychiatric disorders. However, only a small fraction of the adolescents exposed to stress develop psychopathologies [21].

Recently, numerous researchers across many fields have studied adolescence because of its associated intense physical, behavioral, social, and neurologic changes as well as the alarming health statistics relative to this period of life. Researchers have observed that the behavior of adolescents and its related neurological changes bear significant implications for adolescent health. Adolescents are prone to engage in high-risk behaviors with potentially disastrous consequences, including drug abuse, unprotected sex, self-harm, and harm to others, all of which imply a risk of death [22]. According to the 2007 "Youth Risk Behavior Survey" [23], motor-vehicle crashes, unintentional injuries, homicide, and suicide were the most prevalent causes of adolescent mortality, accounting for 72% of the total; all of these causes are preventable. That study suggested that those deaths were partially because of inadequate choices or risky actions (e.g., crashes) and exaggerated emotiveness (e.g., suicides). These findings reinforce the need to understand the biological basis of emotions and promote research regarding adolescent behavior. Nevertheless, the significant role that environmental influences (e.g., reduced parental supervision) play with regard to risks to which adolescents are exposed should not be overlooked, nor should the increased permission that our society provides to adolescents who engage in high-risk situations.

Major Pubertal Changes Due to Hormone Secretion

Two independent, albeit intertwined, processes are responsible for the increase of the sex hormone secretions that occur during peripuberty and puberty: gonadarche and adrenarche. These processes have a similar timing but are controlled by different mechanisms.

Gonadarche is the reactivation of the hypothalamic-pituitary-gonadotropic-gonadal axis. The sizes of the breasts, uterus, and ovaries increases in girls, and the size of the genitals and testicles as well as the amount of hair increases in boys as a result of the increased levels of sex hormones (i.e., estrogens in girls and androgens in boys).

Adrenarche (i.e., pubarche) is the increase of adrenal androgenesis that occurs 2 years before the development of the gonads, usually at 8 years old. Pubic, axillary, and facial hair appears as a result of the increased adrenal production of androgen.

Puberty begins by quickly elevating the production of sex hormones. From 5 to 9 years old, the adrenal glands start producing an increasing amount of androgens, which plays a significant role in the growth of pubic, axillary, and facial hair. A few years later, girls' ovaries increase their production of estrogens, which stimulates the growth of their genitals and the development of their breasts. Among boys, the production of androgens (especially testosterone by the testicles) increases, thereby stimulating the growth of the genitals, body hair, and muscle mass. Importantly, estrogens and androgens are produced in both genders; however, the production of the former is higher in females, and the latter is higher in males. Testosterone promotes the growth of the clitoris, bones, and pubic and axillary hair in females [5].

The need to achieve a "critical weight" to trigger puberty has been postulated; for example, the adipose tissue is a hormonal tissue, and several neuropeptides play a key role in triggering puberty. One of the neuropeptides most widely studied currently is leptin, a protein hormone secreted by the adipose tissue that plays a well-known role in obesity as well as in triggering puberty [24]. The accumulation of leptin in the bloodstream might stimulate the hypothalamus to signal the pituitary gland, which in turn signals the sex glands to increase the corresponding hormone secretion. This pathway might explain why girls with excess weight tend to enter puberty before their thinner counterparts.

The exacerbation of emotions and the mood instability present at the onset of adolescence are associated with hormonal changes [25]. Nevertheless, the influence of other factors such as gender, age, temperament, and the timing of puberty onset should not be overlooked because they can modulate or even surpass the effects of hormones. In any case, aggressiveness among males as well as aggressiveness and depression among females are associated with hormones. Hormones seem to have a stronger relationship with mood states among males and younger individuals, i.e., precisely those who are adjusting to the changes associated with puberty.

The increases in linear growth (height) and body weight known as the adolescent growth spurt is due to the action of the adrenal and gonadal hormones, combined with the increased secretion of growth hormones, somatomedins (i.e., insulin-like growth factor; IGF), and thyroid hormones (which directly act on the bone growth plate). Growth hormone secretion is pulsatile, episodic, and seasonal; furthermore, it tends to occur at night after the first short period of deep sleep. Somatomedins are produced in the liver and act on cell receptors and bone epiphyses; they are carried by "binding" proteins, which are influenced by stress, nutritional, and hormonal factors [5].

All hormonal actions are synergic, and the regulation mechanisms of hormone secretions exhibit a characteristic pulsatile pattern that reflects their mutual interaction as well as the participation of several neurotransmitters and neuropeptides. All of these factors account for the susceptibility of growth to stress, sleep, and fasting as well as to fevers and infections (as is the case for individuals with chronic diseases).

The Hypothalamic-Pituitary-Gonadal System

The hypothalamus and the pituitary gland (or hypophysis) form an interrelated system such that they might be considered a single entity. United with the hypothalamus, the pituitary gland has a double embryonic origin that results in the anterior (or adenohypophysis) and posterior (infundibulum and inferior pituitary stalk) pituitary. The anterior pituitary produces the growth hormone and prolactin. The cells that produce the luteinizing and follicle-stimulating hormones are scattered across the gland. The cells that produce the thyroid-stimulating and adrenocorticotropic (ACTH) hormones are located in the central mucoid wedge, although some ACTH-producing cells are also found in the lateral wings. The pituitary hormones act on various cells and tissues of the human body, which produce other hormones involved in hypothalamus and pituitary self-regulation. Several factors or substances stimulate or inhibit the secretion of hypothalamic-pituitary hormones. Tension, fear, and other emotional stimuli modulate the endocrine system through the connections between the hypothalamus and the higher brain centers via the limbic system. The sleep-wake cycle is a significant physiological modulator that acts via the central nervous system. Several hormones exhibit a circadian pattern of secretion in which their plasma concentrations increase at night after a period of deep sleep [5].

Chemical neurotransmitters such as acetylcholine, gamma-aminobutyric acid, dopamine, histamine, serotonin, and melatonin also interfere with hypothalamic central control. In addition, drugs and other pharmacological substances might act on the neuroendocrine system with a broad range of possible effects. Puberty might be altered or even interrupted following episodes related to the interaction of environmental factors. Therefore, the use of medication or drugs and other environmental factors (e.g., sleep, exercise, emotional disorders, and nutritional status) behave as exogenous factors and have repercussions on the neuroendocrine system during pubertal development. Situations characterized by chronic stress or environmental risk might trigger severe emotional reactions (e.g., fear, anxiety, depression, existential anguish associated with loss, and emotional pain) and stress adaptation responses that might interfere with the neurohormonal mechanisms that promote growth, nutrition, and sexual maturation during puberty.

The human hypothalamic-pituitary-gonadal system differentiates and is active during fetal life and the onset of childhood; its activity is latent for approximately one decade, showing only a low level of activity until puberty when it is reactivated. Thus, puberty likely represents the reactivation of the neurosecretory cells associated with the arcuate nucleus and their endogenous secretion following a long period of minimal activity. Therefore, the onset of puberty is characterized by the activation of the hypothalamic-pituitary-gonadal axis at several levels including gametogenesis, changes in the peripheral receptivity to sex steroids, gonadal steroid secretion, and gonadotropic secretion.

The Hypothalamic-Pituitary-Adrenal System

The maturation of the hypothalamic-pituitary-adrenal axis precedes that of the brain pathways regulating emotions, cognition, and learning such as the prefrontal cortex, hippocampus, amygdala, and ventral striatum [26]. Intense changes in several brain areas contribute to the peculiar behavior of adolescents.

As emphasized above, significant changes in adrenal function occur approximately 2 years before the first hormonal variations of the hypothalamic-pituitary-gonadal axis and gonadal maturation (gonadarche) become detectable.

Adrenarche (i.e., the growth of the androgenic area of the adrenal cortex) and the progressive increase in the serum androgens concentration (first, dehydroepiandrosterone [DHEA] and DHEA-sulfate; second, delta-4-androstenedione [4]) occur during the prepubertal period at approximately 7–8 years old. The onset of either hormonal or physical puberty occurs only 2 years after the onset of the aforementioned processes.

The androgenic action of DHEA and 4 is weak; these hormones promote the growth of pubic hair, mainly among girls (the action of testosterone prevails in boys), and participate in the development of the apocrine and sebaceous glands and bone growth. Importantly, however, these hormones do not play a significant role in the adolescent growth spurt because the greater spurt exhibited by males is due to the action of testosterone. The development of the adrenal cortex is clinically expressed by the increase in the amount and the particular distribution of axillary and pubic hair, the increase in perspiration, and the appearance of body odor and acne.

The conversion of androgens into estrogens occurs in extraglandular tissues under the action of the enzyme aromatase.

The interaction of androgens and estrogens accounts for the development of secondary sex characteristics, ovulation in females, spermatogenesis in males, and the full process of fertilization.

Many adolescents begin their sexual life during early or mid-adolescence. The average age for sexual initiation in Brazil is 15 to 17 years old in females and 13 to 15 years old in males [27]. Factors including the onset of puberty, low self-esteem, having smoked or consumed alcohol, and the absence of excess weight are strongly associated with early sexual debut among females. In the case of males, an active sexual life is associated with having an older age, having been reared with poor parental relationships and weak family ties, and having smoked [28]. Several studies have found an association between sexuality and testosterone production among males, whereas others have emphasized the effects of social factors on pubertal maturation. The increase in testosterone levels is associated with sexual activity. A similar pattern is exhibited among girls, where the elevation of testosterone levels increases the frequency of sexual thoughts and masturbation. Peers exert a striking influence on masturbation as well as on the progression of sexuality and the transition to first sexual intercourse. However, the association between testosterone levels and sexual activity was not found among a population of white girls who regularly attended religious services. The social pressures include messages that target females, either encouraging or restricting the exercise of sexuality, in very different ways compared with males [5].

Therefore, adolescence is a period of physical, emotional, and social vulnerability; adolescents' particular reactivity to stressful situations, cultural factors, hormonal reactions, and brain immaturity are its possible causes.

References

1. Steinberg L. A behavioral scientist looks at the science of adolescent brain development. Brain Cogn. 2010;72:160–4.
2. Vidal J, Mills T, Pang EW, et al. Response inhibition in adults and teenagers: spatiotemporal differences in the prefrontal cortex. Brain Cogn. 2012;79:49–59.
3. Keulers EHH, Stiers P, Jolles J. Developmental changes between ages 13 and 21 years in the extent and magnitude of the BOLD response during decision making. NeuroImage. 2011;54:1442–54.
4. Buck Louis GM, Gray LE, Marcus M, et al. Environmental factors and puberty timing: expert panel research needs. Pediatrics. 2008;121(Suppl. 3):S192–S207.
5. Styne DM, Grumbach MM. Puberty: ontogeny, neuroendocrinology, physiology, and disorders. In: Kronenberg HM, Melmed S, Polonsky KS, et al., editors. Williams textbook of endocrinology. Philadelphia: Elsevier Saunders; 2008. pp. 969–1166.
6. Vitalle MS, Medeiros EH. O adolescente. In: Puccini RF, Hilário MO, editors. Semiologia da criança e do adolescente. Rio de Janeiro: Guanabara Koogan; 2008. pp. 97–108.
7. Brandão ML, Vianna DM, Masson S, et al. Neural organization of different types of fear: implications for the understanding of anxiety. Rev Bras Psiquiatr. 2003;25(Suppl 2):36–41.
8. Canteras NS. Critical analysis of the neural systems organizing innate fear responses. Rev Bras Psiquiatr. 2003;25(Suppl 2):21–4.
9. Carobrez AP. Glutamatergic neurotransmission as molecular target in anxiety. Rev Bras Psiquiatr. 2003;25(Suppl 2):52–8.
10. Schoen-Ferreira TH. A adolescência e a identidade: uma proposta de avaliação e intervenção [doutorado]. São Paulo: Universidade Federal de São Paulo; 2007.
11. Papalia DE, Olds SW, Feldman RD, editors. Desenvolvimento humano. 8th edn. Porto Alegre: Artmed; 2006.
12. Poulton R, Trainor P, Stanton W, et al. The (in)stability of adolescent fears. Behav Res Ther. 1997;35:159–63.
13. Schoen TH, Vitalle MSS. Tenho medo de quê? Rev Paul Ped. 2012;30(1):72–8.
14. Keverne EB. Understanding well-being in the evolutionary context of brain development. Philos Trans R Soc Lond B Biol Sci. 2004;359(1449):1349–58.
15. Luciana M. Adolescent brain development: current themes and future directions. Introduction to the special issue. Brain Cogn. 2010;72:1–5.
16. Sowell ER, Trauner DA, Gamst A, et al. Development of cortical and subcortical brain structures in childhood and adolescence: a structural MRI study. Dev Med Child Neurol. 2002;44(1):4–16.
17. Sowell ER, Thompson PM, Toga AW. Mapping changes in the human cortex throught the span of life. Neuroscientist. 2004;10(4):372–92.
18. Huttenlocher PR. Synaptic density in human frontal cortex-developmental changes and effects of aging. Brain Res. 1979;163(2):195–205.
19. Sowell ER, Delis D, Stiles J, et al. Improved memory functioning and frontal lobe maturation between childhood and adolescence: a structural MRI study. J Int Neuropsychol Soc. 2001;7(3):312–22.
20. Heyman I, Fombonne E, Simmons H, et al. Prevalence of obsessive-compulsive disorder in the British nationwide survey of child mental health. Int Rev Psychiatr. 2003;15(1–2):178–84.
21. Toledo-Rodrigues M, Sandi C. Stress during adolescence increases novelty seeking and risk-taking behavior in male and female rats. Front Behav Neurosci. 2011;5(17):1–10.
22. Sommervile LH, Jones RM, Casey BJ. A time of change: behavioral and neural correlates of adolescent sensitivity to appetitive and aversive environmental cues. Brain Cogn. 2010;72:124–33.
23. Eaton LK, Kann L, Kinchen S, et al. Youth risk behavior surveillance—United States, 2007, Surveillance Summaries. Morb Mort Weekly Rep. 2008;57(SS04):1–131.

24. O'Rahilly S. Life without leptin. Nature. 1998;392:330–1.
25. Buchanan CM, Eccles JS, Becker JB. Are adolescents the victims of raging hormones: evidence for activational effects of hormones on moods and behavior at adolescence. Psychol Bull. 1992;111:62–107.
26. Sowell ER, Thompson PM, Holmes CJ, et al. In vivo evidence for post-adolescent brain maturation in frontal and striatal regions. Nat Neurosci. 1999;2:859–61.
27. Gubert D, Madureira VSF. Iniciação sexual de homens adolescentes. Cienc Saúde Col. 2008;13:2247–56.
28. Garriguet D. Early sexual intercourse. Health Rep. 2005;16(3):9–18.

Maria Sylvia de Souza Vitalle MS in Pediatrics, Federal University of São Paulo/São Paulo School of Medicine—UNIFESP/EPM. PhD in Medicine, UNIFESP/EPM. Associate Professor, Adolescent Medicine Unit, Department of Pediatrics, UNIFESP/EPM. Permanent Lecturer, Graduate Program in Health and Education in Childhood and Adolescence, UNIFESP/EPM. Vice President of the Department of Adolescence, Pediatric Society of São Paulo. Member of the International Association for Adolescence Health. Chief of the Adolescent Medicine Unit, Department of Pediatrics, UNIFESP/EPM.

Chapter 2
Cognitive Development, Learning and Drug Use

Adriana Sampaio, Ana Raquel Mesquita and Óscar Filipe Gonçalves

Use of Cannabis in Adolescence—Epidemiology and Risk Factors

For young adolescents, drugs are usually dichotomously classified as "hard drugs" and "soft drugs". The former category comprises drugs that are known to induce physical and psychological dependence and to pose high risks to the physical and mental health of users. Opiates are perhaps the prototypical example of this category. Soft drugs, which are typically used for recreational purposes, are contrariwise considered by youths as less liable to induce dependency (physical dependence, in particular) and as not posing a significant risk to the users' health. The most typical and generalized example of a "soft drug" is *cannabis*.

The belief in the innocuous nature of *cannabis* causes many youths to deny their use of the drug when asked about their drug habits in the clinical setting; to many youths, the concept of "drug use" is restricted to the so-called synthetic drugs. Instead, the use of *cannabis* is considered an allegedly innocuous behavior that differs little from the use of other substances with various degrees of psychoactive effects (e.g., caffeine and nicotine).

Such beliefs have led *cannabis* to quickly become the most widely used illegal drug among adolescents. For instance, in the United States, following several

Ó. F. Gonçalves (✉) · A. Sampaio · A. R. Mesquita
Laboratory of Neuropsychophysiology, School of Psychology,
University of Minho, Braga, Portugal
e-mail: adriana.sampaio@psi.uminho.pt

A. R. Mesquita
e-mail: ana.mesquita@psi.uminho.pt

Ó. F. Gonçalves
Spaulding Neuromodulation Center, Department of Physical Medicine & Rehabilitation,
Spaulding Rehabilitation Hospital and Massachusetts General Hospital, Harvard Medical School,
Charlestown campus: 79/96 13th Street – Charlestown, MA 02129, USA

Applied Psychology, Bouvé College of Health Sciences,
Northeastern University, 404 Int'l Village, Boston, MA 02115, USA
e-mail: goncalves@psi.uminho.pt

D. De Micheli et al. (eds.), *Drug Abuse in Adolescence*,
DOI 10.1007/978-3-319-17795-3_2

periods of fluctuation in the use of *cannabis* between 1970 and 2008, consumption increased significantly in 2010. Its prevalence was estimated to be 30.4% among adolescents attending the 8th, 10th and 12th grades, with daily consumption levels of 1.2%, 3.3% and 6.1%, respectively [1].

Contrary the extent of its use and the belief in the innocuous nature of *cannabis*, recent concerns have arisen concerning the possibility that continuous exposure to *cannabis* may cause significant psychological damage to adolescents' neurologic and cognitive development during a crucial stage of their psychological development.

There is some evidence that the generalized use of *cannabis* in adolescence represents a significant risk factor for the incidence and persistence of psychotic symptoms. One of the latest studies in this area followed up German youths over 10 years and confirmed that the use of *cannabis* effectively preceded the appearance of psychotic symptoms, even among individuals with no signs of psychotic disorders whatsoever. Thus, the study concluded that *cannabis* represents a significant risk factor for the persistence of such symptoms [2].

Recent evidence also points to the association between continuous use of *cannabis* and the high risk of developing several mood disorders, particularly bipolar and major depressive disorders [3].

Curiously, in a study that followed Australian youths for a 10-year period, even sporadic use of *cannabis* by adolescents was associated with several indicators of psychosocial maladjustment, such as school failure and future use of alcohol and illegal drugs [4].

The mechanisms underlying the psychological and neurobiological development characteristic of adolescence might make adolescents particularly at risk for the use of psychoactive substances such as *cannabis*.

Adolescence is characterized by significant neurobiological maturation processes. With respect to gray matter, the increase in cortical density that takes place before puberty undergoes remarkable reduction in the frontal, temporal and parietal lobes, although the pattern of progression is peculiar to each brain lobe (for instance, the gray matter density reaches its maximum in the frontal lobe at approximately 11 years of age, whereas gray matter density continues to increase in the temporal lobe until 14 years old) [5]. In contrast to the synaptic pruning that results in the reduction of the cortical density, white matter myelination and coherence exhibit significant increases. Of particular interest is the fact that the last areas to attain maturation are the prefrontal regions [6].

The neurobiological development that occurs in adolescence is associated with the maturation of several cognitive functions that are crucial for the future psychological adjustment of youths. Indeed, the structural development of the prefrontal cortex creates the conditions required for greater and more efficient communication with the remainder cortical and subcortical structures, thus endowing adolescents with greater efficiency in decision-making, planning, working memory, meta-cognition, behavior regulation and emotional control [7]. For these reasons, adolescence is typically essential for the maturation of the higher cognitive processes that are crucial for psychological adjustment in adulthood. The possible implications of use of *cannabis* in the development of several psychopathological disorders may be due to an interference with the full process of neurocognitive development. Indeed, one of the hypotheses put forward to account for the pathophysiology of schizophrenia points to a reduced efficiency of the mesocortical dopaminergic system for inhibiting the mesolimbic system. The dopaminergic system, particularly in the prefrontal area, undergoes significant reorganization during adolescence. Thus, any process liable to interfere with the maturation of the mesocortical system might significantly increase the risk of psychotic disorders.

To summarize, *cannabis* is the drug most widely used by adolescents. The use of *cannabis* by adolescents has been identified as a remarkable risk factor for future disorders, particularly psychotic disorders but also affective ones. The neurocognitive changes occurring in adolescence may make this stage of life particularly vulnerable to the use of *cannabis*. Therefore, one of the most immediate effects of *cannabis* may be altered higher cognitive functions, which undergo maturation in adolescence, and a large number of psychiatric disorders may be direct consequences of these alterations.

Next, we analyze the maturation of the endocannabinoid (eCB) system and its implication for cognitive processes. Finally, in the last section of this chapter, we discuss the effects of *cannabis* on cognitive processes in adolescents.

Endocannabinoid System

The eCB system comprises several lipid neuromodulators and their receptors, which participate in various neurophysiological functions, including the processing of pain, memory and mood. At least two cannabinoid receptors (CB), which belong to the seven-transmembrane domain G protein-coupled receptor family, have been characterized: CB1 and CB2 [8, 9]. CB1 is the most abundant CB in the mammal brain and is expressed at high levels in the basal ganglia, hippocampus, cerebellum and cortex [10, 11]. Most of the effects of cannabinoid drugs on the central nervous system (CNS) are mediated by the receptor CB1 [12]. CB2 is mainly located in peripheral sites, particularly in the hematopoietic system [9]. The identification of these receptors followed the discovery of endogenous cannabinoid ligands, among which anandamide and 2-arachidonoylglycerol (2-AG) are the most relevant [13, 14].

eCBs, unlike the classical neurotransmitters, are not stored but are instead released by postsynaptic neurons due to their location in the axon terminals and retrogradely diffuse across the synaptic cleft to stimulate the CB1 receptors on the presynaptic neurons. Activation of CB1 transiently reduces the neurotransmitter release in the presynaptic terminals [15]. Retrograde inhibition of the synaptic transmission was described in GABAergic and glutamatergic synapses throughout the brain, including the neocortex. These findings suggest that eCBs represent a generalized mechanism of synaptic regulation.

In addition, eCB signaling occurs under response [16] and is synapse-specific; eCBs are synthetized from lipid precursors derived from the cell membrane; thus, they are only released when they are required. Together, those features make the eCB system a protective physiological mechanism against excessive stimulation of the receptors to various neurotransmitters, which can easily occur in critical periods of development.

As mentioned previously, adolescence is one of such vulnerable periods of development in which the GABAergic and glutamatergic systems are undergoing remarkable development, particularly in the prefrontal area. The synaptic remodeling that occurs in adolescence is of paramount importance in the refinement of those circuits [17].

Electrophysiological studies indicate that synaptic remodeling is stimulated by neural activity in the form of electrical impulses, which are usually dependent on Ca^{2+} influx. In the presence of eCBs, the intracellular Ca^{2+} level determines the reinforcement or pruning of specific synapses, and the number and quality of such connections are determinant for the improvement of various neural networks [18]. To avoid the excitotoxicity associated with excessive Ca^{2+} influx through the postsynaptic channels in periods of system maturation, synapses should be able to control the amount of glutamate released from the presynaptic terminals. Due to its retrograde action on the GABAergic and glutamatergic synapses, the eCB system plays a crucial role in the regulation of glutamate homeostasis [15, 18, 19].

The use of exogenous cannabinoids in critical periods of development, when the associated processes are particularly intense, might interfere with the regulatory role of the eCB system in GABAergic and glutamatergic neurotransmission [20]. The main psychoactive component of *cannabis* is delta-9-tetrahydrocannabinol (THC), one of the exogenous cannabinoids most widely used by adolescents. THC acts by binding to the presynaptic CB1 receptors in the CNS.

The possible mechanisms of action of *cannabis* include reduction of the activity (loss of binding sites) and desensitization (no longer coupled to G proteins) of the CB1 receptors. These phenomena are consistently observed following chronic administration of synthetic cannabinoids and THC [21]. Thus, exposure to *cannabis*, THC in particular, in adolescence is hypothesized to impair the refining of the neural circuits in the prefrontal cortex (PFC). The dose, the developmental window and the duration of exposure determine the severity and cortical localization of the effects.

Studies in animal models also demonstrate that the PFC dopaminergic system may undergo substantial reorganization during adolescence [22]. Indeed, the dopamine concentration in the PFC decreases after the adolescence peak [23], which occurs concomitant with the improvement in the dopaminergic innervation of the prefrontal pyramidal neurons [24]. In addition, the density of the dopaminergic afferent fibers to the PFC increases during adolescence [25], which differs from other subcortical areas that project to the PFC, such as the striatum, in which the synthesis and turnover of dopamine are lower in adolescence compared to adulthood [23]. The change in the dopamine balance between the PFC and the mesolimbic subcortical structures likely results from the significant pruning of axons that project to the neocortex [26].

The use of exogenous cannabinoids during the abovementioned processes might lead to atypical development of neural circuits in the PFC. The functional implications of such atypical development include aberrant physiological communication between the PFC and other cortical and subcortical structures, mainly as a result of anomalous dopamine and GABA transmission.

The majority of the aforementioned evidence is based on animal studies, typically rodents. In rodents, the eCB system in the prefrontal areas also seems to continue to develop throughout the course of adolescence, which is attended by a dramatic reduction in the cannabinoid binding capacity until the beginning of adulthood [27]. These studies suggest a parallel with cannabinoid use in humans, which also results in cognitive and neurofunctional alterations, as described in the following section.

Neurocognitive Effects of *Cannabis* Use

As stated above, adolescence is characterized by continuous neuronal maturation, which causes increased neurodevelopmental vulnerability to the adverse effects of exposure to exogenous cannabinoids.

This evidence notwithstanding, the neurocognitive effects of *cannabis* following different periods of abstinence have been more widely investigated in adults. The results of these studies are consistent with respect to the association between the chronic use of *cannabis* and acute negative effects on the learning and memory skills, processing speed, executive functioning, decision-making, attention and working memory. However, evidence for the long-term persistence of those neurocognitive deficits is less conclusive.

Significantly fewer studies have been conducted on the acute and chronic effects of use of *cannabis* in adolescence despite the considerable preclinical evidence demonstrating persistent adverse effects following exposure to exogenous cannabinoids in adolescence and that the use of *cannabis* often begins in this period of development.

Neuropsychological studies conducted with adolescents demonstrate acute effects following the use of *cannabis* on measurements of overall intelligence, processing speed, working memory, attention, learning and verbal memory, as well as in executive functions such as planning and sequencing, as well as increased number of perseveration errors [28–30]. A study on long-term effects (after 1 month of supervised abstinence) demonstrated a subtle persistence of neurocognitive deficits relative to the attention skills, processing speed, verbal learning and memory [30, 31]. The results of longitudinal studies are consistent with acute studies and reported a cumulative effect over time, with particularly notably effects on attention skills, processing speed and short- and long-term memory [32, 33].

Overall, neuropsychological studies provide evidence for acute and chronic neurocognitive effects following exposure to exogenous cannabinoids in adolescence. These effects exhibit differential expression according to the age at which use started. Onset of *cannabis* use before age 16–17 years old is a strong predictor of impairment of the attention skills [34] and reduction of verbal IQ [31].

Several studies demonstrate parallel evidence that the deleterious effects of *cannabis* on neurocognition are attended by changes in brain structure and function. The results of structural magnetic resonance imaging (MRI) are inconsistent in the detection of global or regional changes in adults [35, 36]; however, MRI studies conducted with adolescents clearly demonstrate significant neurobiological and neurofunctional effects following the use of *cannabis*. Wilson et al. [37] reported that the onset of *cannabis* use in adolescence was associated with a reduction in the gray matter percentage and an increase in the white matter percentage after normalization by the total intracranial volume. The volumetric change in white matter is consistent with alterations in the integrity and structure of several of its bundles, including changes in the fractional anisotropy and mean diffusivity of the corpus callosum [38], fronto-parietal [39] and fronto-temporal circuits [40].

Functional alterations of these circuits have also been documented by studies using functional MRI and include abnormalities in the patterns of frontal, temporal and parietal activation among adolescent *cannabis* users (both acute effects and effects following different periods of abstinence) with respect to visual attention [41], inhibitory processing [42], and verbal [43] and spatial [44] working memory tasks.

Some studies have reported a connection between structural and functional aspects in verbal memory tasks, particularly, a bilateral reduction of the hippocampal formation volume in individuals who made frequent use of *cannabis* [45]. Moreover, unlike the control group, a positive association between volume and performance in verbal memory tasks was not observed in the group of *cannabis* users. More specifically, studies that performed functional neuroimaging during the performance of verbal and spatial working memory tasks reported increased activation in a brain network that included the parietal cortex, hippocampus, and anterior cingulate cortex in conjunction with a reduction of the BOLD response in the dorsolateral PFC and occipital lobe. The PFC was differentially activated following recent *cannabis* use (compared with the abstinence condition). In particular, an increased brain response was observed in the PFC, right superior parietal cortex, and bilateral insula [46]. These findings indicate different acute and chronic effects of cannabis use on the brain function of adolescents, suggesting that recent *cannabis* users require greater recruitment of the brain areas underlying the working memory circuits, whereas the abstinent adolescents exhibit mechanisms of brain reorganization and plasticity, particularly in the PFC. The pattern of increased cortical activation seems to be specific to the type of cognitive task under assessment. In attention tasks (inhibitory processing), a marked BOLD response was observed in the parietal lobe in conjunction with activation of the dorsolateral PFC.

The changes observed in the brain areas responsible for the various neurocognitive tasks (visual attention, inhibitory processing, verbal and spatial working memory), such as the frontal lobe, hippocampus, basal ganglia and striatum, might be related to the fact that these brain regions are rich in cannabinoid receptors and thus more susceptible to the action of THC. Finally, the magnitude of these changes seems to correlate with early exposure to exogenous cannabinoids in adolescence [43].

To summarize, the results of functional studies seem to reflect changes in the neural circuits associated with specific cognitive domains in adolescent *cannabis* users. An impact on the brain structure as well as consistent short-term changes in neurocognitive functioning have been observed. Together, those studies point to the presence of neurotoxic and functional effects following exposure to *cannabis* in adolescence, with long-lasting and persistent impact on the prefrontal and parietal neuronal networks.

Indeed, as mentioned above, the concentration of CB1 receptors increases during adolescence in some brain areas, such as the PFC. Thus, the use of exogenous cannabinoids affects the maturation processes (for instance, reduction of the activity and desensitization of the CB1 receptors via synaptic regulation) that occur during the course of adolescence, with functional implications for the synaptic plasticity that underlies learning and memory processes. Together, these studies indicate that frequent use of *cannabis* in adolescence exerts a negative influence on neuromaturation and on corresponding alterations in neurocognitive functioning.

Conclusion

The study of the neurobiological, cognitive and behavioral effects of *cannabis* use is complex due to the large number of variables that must be methodologically taken into consideration. Indeed, comparison among studies is difficult due to the varying levels of exposure to *cannabis*, polydrug use, and the substantial comorbidity with psychiatric disorders, which hinder attempts to establish whether the impact of *cannabis* on neurocognitive functioning is exclusively due to the use of *cannabis*.

In addition, limitations relative to the recruitment of volunteers (treatment/follow up versus non-treatment), the distinction between acute or subacute effects, the abstinence syndromes, the definition of frequent versus sporadic users, the assessment of the influence of social and educational opportunities, and premorbid cognitive function further increase the complexity inherent to the study of the effects of *cannabis* use.

Future studies must take these variables into consideration to provide a more thorough understanding of the adverse effects of exposure to exogenous cannabinoids by means of multimodal and translational integrated approaches using neuroimaging and neurocognitive techniques as well as animal models. This approach will also allow a holistic view of the phenomenon, as well as a critical analysis of the possible therapeutic implications of the use of *cannabis*.

References

1. Johnston LD, O'Malley PM, Bachman JG, et al. Monitoring the future national results on adolescent drug use: overview of key findings, 2010. Ann Arbor: Institute for Social Research, The University of Michigan; 2011.
2. Kuepper R, van Os J, Lieb R, et al. Continued cannabis use and risk of incidence and persistence of psychotic symptoms: 10 year follow-up cohort study. BMJ. 2011;342:738.
3. Sanches RF, Marques JM. Cannabis and mood. Rev Bras Psiquiatr. 2010;32(2):173–80.
4. Degenhardt L, Coffey C, Carlin JB, et al. Outcomes of occasional cannabis use in adolescence: 10-year follow-up study in Victoria, Australia. Br J Psychiatry. 2010;196:290–5.
5. Gogtay N, Thompson PM. Mapping gray matter development: implications for typical development and vulnerability to psychopathology. Brain Cogn. 2010;72(1):6–15.

6. Schmithorst VJ, Yuan W. White matter development during adolescence as shown by diffusion MRI. Brain Cogn. 2010;72(1):16–25.
7. Steinberg L. Cognitive and affective development in adolescence. Trends Cogn Sci. 2005;9(2):69–74.
8. Matsuda LA, Lolait SJ, Brownstein MJ, et al. Structure of a cannabinoid receptor and functional expression of the cloned cDNA. Nature. 1990;346(6284):561–4.
9. Munro S, Thomas KL, Abu-Shaar M. Molecular characterization of a peripheral receptor for cannabinoids. Nature. 1993;365(6441):61–5.
10. Herkenham M, Lynn AB, Johnson MR, et al. Characterization and localization of cannabinoid receptors in rat brain: a quantitative in vitro autoradiographic study. J Neurosci. 1991;11(2):563–83.
11. Glass M, Dragunow M, Faull RL. Cannabinoid receptors in the human brain: a detailed anatomical and quantitative autoradiographic study in the fetal, neonatal and adult human brain. Neuroscience. 1997;77(2):299–318.
12. Huestis MA, Gorelick DA, Heishman SJ, et al. Blockade of effects of smoked marijuana by the CB1-selective cannabinoid receptor antagonist SR141716. Arch Gen Psychiatry. 2001;58(4):322–8.
13. Devane WA, Hanus L, Breuer A, et al. Isolation and structure of a brain constituent that binds to the cannabinoid receptor. Science. 1992;258(5090):1946–9.
14. Sugiura T, Kondo S, Sukagawa A, et al. 2-Arachidonoylglycerol: a possible endogenous cannabinoid receptor ligand in brain. Biochem Biophys Res Commun. 1995;215(1):89–97.
15. Wilson RI, Nicoll RA. Endocannabinoid signaling in the brain. Science. 2002;296(5568):678–82.
16. Piomelli D. The molecular logic of endocannabinoid signalling. Nat Rev Neurosci. 2003;4(11):873–84.
17. Citri A, Malenka RC. Synaptic plasticity: multiple forms, functions, and mechanisms. Neuropsychopharmacology. 2008;33(1):18–41.
18. Chevaleyre V, Takahashi KA, Castillo PE. Endocannabinoid-mediated synaptic plasticity in the CNS. Annu Rev Neurosci. 2006;29:37–76.
19. Schlicker E, Kathmann M. Modulation of transmitter release via presynaptic cannabinoid receptors. Trends Pharmacol Sci. 2001;22(11):565–72.
20. Mato S, Chevaleyre V, Robbe D, et al. A single in-vivo exposure to delta 9THC blocks endocannabinoid-mediated synaptic plasticity. Nat Neurosci. 2004;7(6):585–6.
21. Sim-Selley LJ, Martin BR. Effect of chronic administration of R-(+)-[2,3-Dihydro-5-methyl-3-[(morpholinyl)methyl]pyrrolo[1,2,3-de]- 1,4-b enzoxazinyl]-(1-naphthalenyl)methanone mesylate (WIN55,212–2) or delta(9)-tetrahydrocannabinol on cannabinoid receptor adaptation in mice. J Pharmacol Exp Ther. 2002;303(1):36–44.
22. Spear LP. The adolescent brain and age-related behavioral manifestations. Neurosci Biobehav Rev. 2000;24(4):417–63.
23. Andersen SL, Rutstein M, Benzo JM, et al. Sex differences in dopamine receptor overproduction and elimination. Neuroreport. 1997;8(6):1495–8.
24. Tseng KY, O'Donnell P. D2 dopamine receptors recruit a GABA component for their attenuation of excitatory synaptic transmission in the adult rat prefrontal cortex. Synapse. 2007;61(10):843–50.
25. Rosenberg DR, Lewis DA. Changes in the dopaminergic innervation of monkey prefrontal cortex during late postnatal development: a tyrosine hydroxylase immunohistochemical study. Biol Psychiatry. 1994;36(4):272–7.
26. Woo TU, Pucak ML, Kye CH, et al. Peripubertal refinement of the intrinsic and associational circuitry in monkey prefrontal cortex. Neuroscience. 1997;80(4):1149–58.
27. Belue RC, Howlett AC, Westlake TM, et al. The ontogeny of cannabinoid receptors in the brain of postnatal and aging rats. Neurotoxicol Teratol. 1995;17(1):25–30.
28. Lane SD, Cherek DR, Tcheremissine OV, et al. Response perseveration and adaptation in heavy marijuana-smoking adolescents. Addict Behav. 2007;32(5):977–90.
29. Harvey MA, Sellman JD, Porter RJ, et al. The relationship between non-acute adolescent cannabis use and cognition. Drug Alcohol Rev. 2007;26(3):309–19.

30. Medina KL, Nagel BJ, Park A, et al. Depressive symptoms in adolescents: associations with white matter volume and marijuana use. J Child Psychol Psychiatry. 2007;48(6):592–600.
31. Pope HG Jr., Gruber AJ, Hudson JI, et al. Early-onset cannabis use and cognitive deficits: what is the nature of the association? Drug Alcohol Depend. 2003;69(3):303–10.
32. Tapert SF, Granholm E, Leedy NG, et al. Substance use and withdrawal: neuropsychological functioning over 8 years in youth. J Int Neuropsychol Soc. 2002;8(7):873–83.
33. Fried PA, Watkinson B, Gray R. Neurocognitive consequences of marihuana: a comparison with pre-drug performance. Neurotoxicol Teratol. 2005;27(2):231–9.
34. Ehrenreich H, Rinn T, Kunert HJ, et al. Specific attentional dysfunction in adults following early start of cannabis use. Psychopharmacology (Berl). 1999;142(3):295–301.
35. Block RI, O'Leary DS, Ehrhardt JC, et al. Effects of frequent marijuana use on brain tissue volume and composition. Neuroreport. 2000;11(3):491–6.
36. Yucel M, Solowij N, Respondek C, et al. Regional brain abnormalities associated with long-term heavy cannabis use. Arch Gen Psychiatry. 2008;65(6):694–701.
37. Wilson W, Mathew R, Turkington T, et al. Brain morphological changes and early marijuana use: a magnetic resonance and positron emission tomography study. J Addict Dis. 2000;19(1):1–22.
38. Arnone D, Barrick TR, Chengappa S, et al. Corpus callosum damage in heavy marijuana use: preliminary evidence from diffusion tensor tractography and tract-based spatial statistics. Neuroimage. 2008;41(3):1067–74.
39. Bava S, Frank LR, McQueeny T, et al. Altered white matter microstructure in adolescent substance users. Psychiatry Res. 2009;173(3):228–37.
40. Ashtari M, Cervellione K, Cottone J, et al. Diffusion abnormalities in adolescents and young adults with a history of heavy cannabis use. J Psychiatr Res. 2009;43(3):189–204.
41. Chang L, Yakupov R, Cloak C, et al. Marijuana use is associated with a reorganized visual-attention network and cerebellar hypoactivation. Brain. 2006;129(Pt 5):1096–112.
42. Abdullaev Y, Posner MI, Nunnally R, et al. Functional MRI evidence for inefficient attentional control in adolescent chronic cannabis abuse. Behav Brain Res. 2010;215(1):45–57.
43. Becker B, Wagner D, Gouzoulis-Mayfrank E, et al. The impact of early-onset cannabis use on functional brain correlates of working memory. Prog Neuropsychopharmacol Biol Psychiatry. 2010;34(6):837–45.
44. Schweinsburg AD, Schweinsburg BC, Cheung EH, et al. fMRI response to spatial working memory in adolescents with comorbid marijuana and alcohol use disorders. Drug Alcohol Depend. 2005;79(2):201–10.
45. Ashtari M, Avants B, Cyckowski L, et al. Medial temporal structures and memory functions in adolescents with heavy cannabis use. J Psychiatr Res. 2011;45(8):1055–66.
46. Schweinsburg AD, Schweinsburg BC, Medina KL, et al. The influence of recency of use on fMRI response during spatial working memory in adolescent marijuana users. J Psychoactive Drugs. 2010;42(3):401–12.

Adriana Sampaio PhD in Psychology, University of Minho, Portugal.

Ana Raquel Mesquita BS in Applied Biology and PhD in Health, Biological and Biomedical Sciences, University of Minho, Portugal. Assistant investigator, Laboratory of Neuropsychophysiology, School of Psychology.

Óscar Filipe Gonçalves Full Professor of Psychology, University of Minho, Portugal. BA in Psychology, University of Porto, Portugal. PhD in Psychotherapy and Psychological Counseling, University of Massachusetts, USA. Director, Laboratory of Neuropsychophysiology; President, School of Psychology, University of Minho, Portugal.

Chapter 3
Neural Development in Adolescence

André Luiz Monezi Andrade, Denise De Micheli, Eroy Aparecida da Silva, Maria Lucia Oliveira de Souza-Formigoni and Francine de Oliveira Goeldner

Introduction

The development of the central nervous system involves several stages of maturation according to age. Several authors asserted that although human beings are born with a huge number of neurons approximately half of them are lost within the two first years of life [1]. The rationale underlying this process is that the brain does not need an excessive number of nerve cells but need to amplify the already existing connections, increasing the number of synapses. In other words, during the course of development, the nervous system prioritizes qualitative aspects (improved transmission among and integration of neurons via synapses) over quantitative aspects (a large number of neurons with rather isolated actions). Recently, Brazilian

A. L. M. Andrade (✉)
Department of Psychology, Anhembi Morumbi University—*Laureate International Universities®*, São Paulo, Brazil
e-mail: andremonezi@gmail.com

D. De Micheli
Department of Psychobiology, Universidade Federal de São Paulo—UNIFESP, São Paulo, Brazil
e-mail: demicheli.unifesp@gmail.com

E. A. da Silva
Associação Fundo de Incentivo à Pesquisa (AFIP), São Paulo, Brazil
e-mail: eroyntc@gmail.com

M. L. O. de Souza-Formigoni
Department of Psychobiology, Escola Paulista de Medicina, Universidade Federal de São Paulo—UNIFESP, São Paulo, Brazil
e-mail: mlformig@gmail.com

F. de Oliveira Goeldner
Curitiba, Brazil
e-mail: francinegoeldner@hotmail.com

D. De Micheli et al. (eds.), *Drug Abuse in Adolescence*,
DOI 10.1007/978-3-319-17795-3_3

researchers elaborated new techniques to count neurons [2, 3] and found that their number is approximately 86 billion in healthy human beings [4]. In certain brain areas, neurogenesis continues throughout life, particularly in the case of the hippocampus [5, 6], a memory and learning critical area. That process is exacerbated during adolescence. Some authors have found higher levels of hippocampal neurogenesis in adolescent mice than in adults [7]. In addition, this process seems to be highly sensitive to the action of drugs, such as alcohol [8, 9].

During adolescence, feelings, and novel experiences are mediated by areas included in the limbic system [10, 11]. The group of researchers coordinated by Monique Ernst, who contributed a chapter to this book, developed a model (triadic model) that allows for the understanding of the neural mechanisms that make adolescents more vulnerable to risk behaviors [12–14]. Overall, this model suggests that imbalance in the integration of the three main neural systems is strongly associated with risk behaviors. The first system, the neural reward system (commonly known as the pleasure center), consists of projections from dopaminergic neurons in the ventral tegmental area (VTA) to the nucleus accumbens (NAc) (where dopamine is released, providing reinforcing effects). The second system integrates the amygdala and its connections. The fact that adolescents are less cautious in the assessment of the risk posed by their behavior compared to adults is associated with imbalance among the circuits corresponding to this second system [15]. The main component of the third system is the prefrontal cortex (PFC), which is traditionally known to participate in cognitive functions (decision-making, working memory, attention and perception, among others). The maturation of the PFC and other areas, such as the medial and the ventral PFC, is delayed in adolescents [16, 17]. The triadic model suggests that in adolescents, such alterations are involved in decision-making, influencing their behavior and making them more vulnerable to choices more valued in the short run.

The vulnerability of adolescents to substance use is supported not only by changes in the brain structure but also by several neurotransmitter systems, among which dopaminergic, serotonergic, noradrenergic, and glutamatergic systems stand out. We discussed each of these systems further, together with the main corresponding changes that occur during adolescence.

Dopamine

Dopamine (DA) is produced in the VTA [18], in the substantia nigra [19,] and also, in a smaller proportion, in the tuberoinfundibular area [20]. The dopaminergic system has a critical role in attention [21], motor system [22], hormone regulation [23], and sleep-wake cycle mechanisms [24].

Many studies pointed out the involvement of DA in substance abuse and dependence [25]. The prevalence of chemical dependence is higher among adolescents with a history of mental disorders (mainly schizophrenia, borderline personality disorder, attention deficit hyperactivity disorder, and mood disorders) compared to adolescents without such history [26, 27]. Curiously, an association between disorders of the dopaminergic system and the above mentioned conditions was observed, which suggests that DA is relevant in the regulation of risk behaviors [28, 29].

The reward system seems to be altered in adolescence compared to other periods of life. High production of dopaminergic receptors was detected in the NAc of adolescent rats; those animals also exhibited greater sensitivity compared to the adults [30–32]. Contrariwise, a low concentration of dopaminergic synapses was found in the ventral striatum, in addition to faster reception of DA in that area [33, 34]. Taken together, these findings suggest that DA release in the NAc can promote a hyperactivation of the reward system, resulting in a much greater reinforcing effect in adolescent rats than in adults [35]. The rationale underlying that theory is that high DA levels are involved in the reinforcing effects of behaviors presented by organisms. Therefore, when adolescents undergo a few pleasurable experiences (low DA release in NAc), they continue to seek new reinforcing experiences (increased DA release in the NAc) [25]. Consequently, such experiences might acquire a special character because the adolescents might repeat them several times. This hypothesis seems to agree with some findings reported in the literature that show that DA release in the reward system in adolescent rats is greater when they are exposed to new situations [36] or given cocaine [37], methamphetamine [38], or alcohol [39, 40].

The role of DA in risk behaviors is not limited to the areas traditionally associated with emotional aspects. The PFC is subject to the influence of the dopaminergic system. A group of researchers from the University of South Carolina chaired by professor Peter Kalivas (who authored a chapter in this book) formulated a theory to explain how DA-mediated inhibition of PFC might influence the preference for behaviors associated with drug seeking at the expense of other activities that used to be a part of the individual's routine [41, 42]. According to those authors, chemical dependence is mediated by neurochemical alterations that occur in three stages. The third and final stage, when dependence is fully consolidated, may be characterized by permanent alteration (hyperactivation) of the neurotransmission from the PFC neurons to the NAc.

The consequence of the excessive activation of the projections from the PFC to the NAc is a symptom that commonly occurs among individuals who abstain from a drug known as craving, which is characterized by an intense and permanent desire to use the corresponding substance. Summarily (for further details, see the corresponding chapter), PFC sends projections to the NAc using glutamate (which is considered to be main excitatory neurotransmitter in the central nervous system). Glutamate release in the NAc is mediated by DA receptors D1 and D2, and thus, the actual amount of DA that is released depends on the receptor subtype that is activated. As a rule, the activation of receptor D2 is related to reduced synaptic inhibition, while the activation of receptor D1 is related to increased synaptic inhibition. In other words, the hyperactivation of receptor D2 in the PFC permits access to various excitatory stimuli that might reach the NAc. Contrariwise, the hyperactivation of receptor D1 permits access to few excitatory stimuli that project to the NAc. To summarize, the hyperactivation of receptor D1 forces different stimuli to compete to be projected to the NAc [43]. According to this theory, in abstinent drug-dependent individuals, the D1 receptors are hyperactivated, and thus, only the behaviors associated with the acquisition and use of the drug become a priority. That hypothesis allows us understand why the behavior of many people who used to perform several reinforcing tasks (balance in the D1 and D2 receptor activation; e.g., going to the movies, going out, meeting with friends, having lunch with family, exercising) becomes restricted to seeking drugs (e.g., friends associated with drug use, quest for places and situations where they might find the substance).

Norepinephrine

Norepinephrine (NE), or noradrenalin, is a monoamine produced from the amino acid tyrosine (by means of the enzyme dopamine-hydroxylase), which occurs in a brainstem area called the locus *coeruleus*. NE was one of the first neurotransmitters to be associated with depression because the tricyclic antidepressants (e.g., imipramine) exhibit a high ability to reuptake NE [44]. Release of NE in the PFC exerts a significant influence on decision-making [45] in addition to having a strong association with attention deficit hyperactivity disorder (ADHD) [46]. Methylphenidate is an amphetamine that increases the DA and NE levels in areas such as the PFC, thus increasing the attention span. Atomoxetine, which acts almost exclusively on the noradrenergic system, has recently been made available in some countries for the treatment of ADHD; there is strong evidence for its effectiveness to improve the core symptoms of ADHD [47].

Although there has been evidence for a strong relationship between NE and the reward system for more than 40 years [48], NE has become a focus of particular attention only in the past decades as a possible target for the treatment of chemical dependence. Researchers have known since the 1970s that in rats trained to self-administer cocaine, the frequency of that behavior does not increase following lesions of the noradrenergic system [49]. Stimulants increase the levels of monoamines, including NE, in several brain areas. Although the role DA plays in the reward system was more thoroughly investigated, NE also seems to have a relevant function involving the indirect control of DA release in the NAc. Some researchers have suggested that the function of NE is similar to the function of serotonin, depending on the substance used. For example, the affinity of cocaine for NE, serotonin, and NE transporters is similar, whereas the affinity of methamphetamine for the NE transporters is nine times greater [50, 51]. In addition, the stimulating effect of amphetamines seems to be more related to NE [52, 53].

Certain authors believe that the noradrenergic system is only completed by the end of adolescence, while the maturation of the serotonergic system occurs earlier [54]. As the synthesis of NE depends on the production of DA, the corresponding systems are intrinsically associated, which bears significant implications for therapeutic decision-making. For example, the therapeutic response of children and adolescents to tricyclic antidepressants is not satisfactory compared to selective serotonin reuptake inhibitors [55]. A greater density of binding to NE transporters was detected in adolescent rats compared to adults [56], most likely due to the larger number of synapses that recruit NE. In adolescent rats given low doses of acetaldehyde and provided with increasing doses of nicotine, higher noradrenergic activity was found in the NAc than in adult ones [57]. In addition, the authors of that study assessed the same pattern of response in six other areas and concluded that NE stimulation seems to be age-dependent.

Serotonin

Serotonin, or 5-hydroxitriptamine (5-HT), is a neurotransmitter with a broad scope of action throughout the central nervous system and the peripheral nervous system. It acts on 15 subtypes of receptors, which are classified into 7 classes and 12 subclasses: 5-HT1 (5-$HT1_A$, 5-$HT1_B$, 5-$HT1_D$, 5-$HT1_E$ and 5-$HT1_F$), 5-HT2 (5-$HT2_A$, 5-$HT2_B$ and 5-$HT2_C$), 5-HT3 (5-$HT3_A$, 5-$HT3_B$), 5-HT4, 5-HT5 (5-$HT5_A$ and 5-$HT5_B$), 5-HT6 and 5-HT7. On the grounds of the large amount and wide distribution of those receptors, it is easy to understand the critical role serotonin plays in the modulation of several physiological and behavioral mechanisms, among which the following stand out: mood regulation [58], sleep-wake cycle (sleep induction in particular [59]), vasoconstriction [60], chronic pain [61], dietary intake [62], and sexual behavior [63].

Serotonin was discovered by Vittorio Erspamer, an Italian researcher at the University of Parma. In 1935, Erspamer observed a contraction of bowel tissue in response to an extract composed of enterochromaffin cells (cells that line the gastrointestinal tract) and called that substance enteramine. Later, in 1948, Maurice Rapport detected that same substance in the plasma (the platelets in particular) and named it serotonin [64]. Serotonin is mainly produced in the raphe nuclei, which are located in the brainstem and certain areas of the gastrointestinal tract.

In addition to some of the functions described here, serotonin plays a critical role in the modulation of the PFC neurons. Overall, the PFC neurons are activated by dopaminergic projections and inhibited by serotonergic projections [65]. Individuals with low serotonin levels are more aggressive and impulsive compared to individuals with normal serotonin levels [66]. Relative to adolescents, the findings by Lambe et al. [67] in adolescent Rhesus monkeys are particularly relevant. Those authors found that the dopaminergic pathway was approximately three times more active than the serotonergic pathway and that the level of the DA precursor in the PFC was higher than serotonin precursors. Those data suggest that the frequency of impulsive and aggressive behavior is higher in adolescents than in adults.

The action of serotonin in the reward system seems to be different in adolescence. The turnover rate of serotonin in the NAc is approximately four times higher in adolescent rats compared to adults [68]. In addition, an association was found between low serotonergic activity in adolescents and behaviors characteristic of certain anxiety disorders and high alcohol consumption [69]. Those data are consistent with the finding that adolescent rats that drink high amounts of alcohol exhibit a long-lasting increase of serotonin transporter levels in the limbic system.

Several authors have suggested that the hallucinations induced by various types of drugs are mainly mediated by the hyperactivation of 5-HT2_A receptors [70]. The concentration of those receptors is high in cortical areas at the onset of adolescence and decreases as one ages [71]. Therefore, the hallucinogen effects might possibly be stronger in adolescents than in adults.

Glutamate

Glutamate (Glu) is the main excitatory neurotransmitter in the central nervous system; it binds to 11 receptors, three ionotropic (which are the most widely studied relative to chemical dependence), namely the NMDA, AMPA, and kaianate receptors, and eight metabotropic receptors (mGluRs$_{1-8}$). As the number of receptors is large and Glu is produced in several areas of the central nervous system, the action of this neurotransmitter is quite diffuse and complex. It has paramount importance in the development of new synaptic connections (long-term potential), memory, learning, attention and mental disorders, among others. In addition, learning processes related to the use of substances occur via the formation of new synapses, involving Glu as the main neurotransmitter. The chronic or acute use of drugs of abuse seems to induce modifications in several neurochemical processes in the glutamatergic receptors, and therefore, the best evidence available is provided by stimulant drugs [72].

The glutamatergic receptors seem to modulate DA release in the VTA as well as GABAergic projections from the NAc to the VTA. Alcohol blocks the action of Glu by behaving as an antagonist to the glycine-binding site of glutamatergic receptors, such as NMDA. The lack of glycine binding to the receptor is related with the absence of excitatory effects.

The increase in the expression of NMDA and dopaminergic receptors D_1, D_2 and D_4 [30] as adolescence progresses results in greater DA synthesis in the NAc and increased density of the DA transporter [73].

Regarding the role of Glu in chemical dependence, consistent evidence indicates that Glu elevation is related to craving (intense desire for a drug) via glutamatergic projections from the PFC to the NAc. Acamprosate is one of the agents that have been used recently with some success to treat cravings for alcohol, as it is an antagonist of glutamatergic receptors [74].

Figure 3.1 summarizes the topics discussed in this chapter and depicts certain characteristics of the neurotransmitter systems described above.

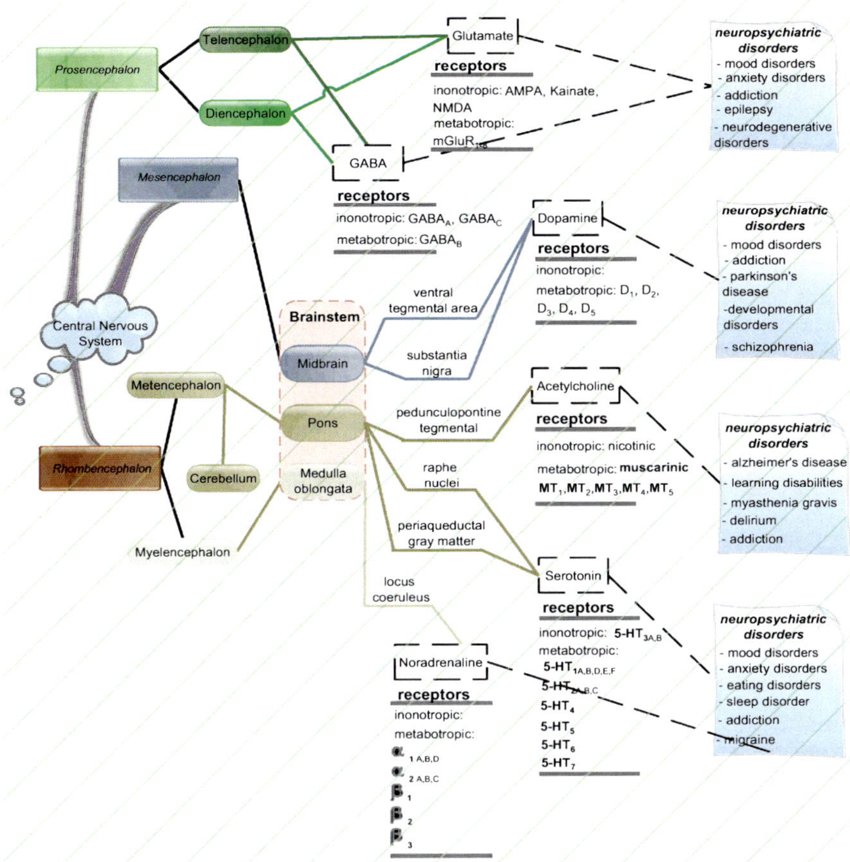

Fig. 3.1 Distribution of receptors of the main neurotransmitters modulated by psychotropic drugs and their relationship with behavior

References

1. Wilke M, Holland SK. Variability of gray and white matter during normal development: a voxel-based MRI analysis. Neuroreport. 2003;14(15):1887–90.
2. Herculano-Houzel S, Lent R. Isotropic fractionator: a simple, rapid method for the quantification of total cell and neuron numbers in the brain. J Neurosci. 2005;25(10):2518–21.
3. Herculano-Houzel S, Ribeiro P, Campos L, et al. Updated neuronal scaling rules for the brains of glires (rodents/lagomorphs). Brain Behav Evol. 2011;78(4):302–14.
4. Azevedo FA, Carvalho LR, Grinberg LT, et al. Equal numbers of neuronal and nonneuronal cells make the human brain an isometrically scaled-up primate brain. J Comp Neurol. 2009;513(5):532–41.
5. Bruel-Jungerman E, Rampon C, Laroche S. Adult hippocampal neurogenesis, synaptic plasticity and memory: facts and hypotheses. Rev Neurosci. 2007;18(2):93–114.
6. Toni N, Sultan S. Synapse formation on adult-born hippocampal neurons. Eur J Neurosci. 2011;33(6):1062–8.

7. He J, Crews FT. Neurogenesis decreases during brain maturation from adolescence to adulthood. Pharmacol Biochem Behav. 2007;86(2):327–33.
8. Crews FT, Mdzinarishvili A, Kim D, et al. Neurogenesis in adolescent brain is potently inhibited by ethanol. Neuroscience. 2006;137(2):437–45.
9. Crews FT, Nixon K. Mechanisms of neurodegeneration and regeneration in alcoholism. Alcohol Alcohol. 2009;44(2):115–27.
10. Alcaro A, Huber R, Panksepp J. Behavioral functions of the mesolimbic dopaminergic system: an affective neuroethological perspective. Brain Res Rev. 2007;56(2):283–321.
11. Alcaro A, Panksepp J. The SEEKING mind: primal neuro-affective substrates for appetitive incentive states and their pathological dynamics in addictions and depression. Neurosci Biobehav Rev. 2011;35(9):1805–20.
12. Ernst M, Fudge JL. A developmental neurobiological model of motivated behavior: anatomy, connectivity and ontogeny of the triadic nodes. Neurosci Biobehav Rev. 2009;33(3):367–82.
13. Ernst M, Pine DS, Hardin M. Triadic model of the neurobiology of motivated behavior in adolescence. Psychol Med. 2006;36(3):299–312.
14. Ernst M, Romeo RD, Andersen SL. Neurobiology of the development of motivated behaviors in adolescence: a window into a neural systems model. Pharmacol Biochem Behav. 2009;93(3):199–211.
15. Bava S, Tapert SF. Adolescent brain development and the risk for alcohol and other drug problems. Neuropsychol Rev. 2010;20(4):398–413.
16. Chan T, Kyere K, Davis BR, et al. The role of the medial prefrontal cortex in innate fear regulation in infants, juveniles, and adolescents. J Neurosci. 2011;31(13):4991–9.
17. Forbes EE, Ryan ND, Phillips ML, et al. Healthy adolescents' neural response to reward: associations with puberty, positive affect, and depressive symptoms. J Am Acad Child Adolesc Psychiatry. 2010;49(2):162–72 e161–5.
18. Mercuri NB, Calabresi P, Bernardi G. The electrophysiological actions of dopamine and dopaminergic drugs on neurons of the substantia nigra pars compacta and ventral tegmental area. Life Sci. 1992;51(10):711–8.
19. Spano PF, Di Chiara G, Tonon GC, et al. A dopamine-stimulated adenylate cyclase in rat substantia nigra. J Neurochem. 1976;27(6):1565–8.
20. Wedzony K, Czyrak A. Presence and function of dopamine D1 receptors in the rat ventral tegmental area. Pol J Pharmacol. 1997;49(5):277–81.
21. Silva Jr N, Szobot CM, Anselmi CE, et al. Attention deficit/hyperactivity disorder: is there a correlation between dopamine transporter density and cerebral blood flow? Clin Nucl Med. 2011;36(8):656–60.
22. Sealfon SC. Dopamine receptors and locomotor responses: molecular aspects. Ann Neurol. 2000;47(4 Suppl 1):S12–19.
23. Reymond MJ, Porter JC. Involvement of hypothalamic dopamine in the regulation of prolactin secretion. Horm Res. 1985;22(3):142–52.
24. Monti JM. Catecholamines and the sleep-wake cycle. I. EEG and behavioral arousal. Life Sci. 1982;30(14):1145–57.
25. Di Chiara G. Drug addiction as dopamine-dependent associative learning disorder. Eur J Pharmacol. 1999;375(1–3):13–30.
26. Bucholz KK. Nosology and epidemiology of addictive disorders and their comorbidity. Psychiatr Clin North Am. 1999;22(2):221–40.
27. Compton WM, Thomas YF, Stinson FS, et al. Prevalence, correlates, disability, and comorbidity of DSM-IV drug abuse and dependence in the United States: results from the national epidemiologic survey on alcohol and related conditions. Arch Gen Psychiatry. 2007;64(5):566–76.
28. Jimenez-Castro L, Hare E, Medina R, et al. Substance use disorder comorbidity with schizophrenia in families of Mexican and Central American ancestry. Schizophr Res. 2010;120(1–3):87–94.

29. Qi Z, Kikuchi S, Tretter F, et al. Effects of dopamine and glutamate on synaptic plasticity: a computational modeling approach for drug abuse as comorbidity in mood disorders. Pharmacopsychiatry. 2011;44(Suppl 1):S62–S75.
30. Tarazi FI, Baldessarini RJ. Comparative postnatal development of dopamine D(1), D(2) and D(4) receptors in rat forebrain. Int J Dev Neurosci. 2000;18(1):29–37.
31. Brenhouse HC, Sonntag KC, Andersen SL. Transient D1 dopamine receptor expression on prefrontal cortex projection neurons: relationship to enhanced motivational salience of drug cues in adolescence. J Neurosci. 2008;28(10):2375–82.
32. Camarini R, Griffin 3rd WC, Yanke AB, et al. Effects of adolescent exposure to cocaine on locomotor activity and extracellular dopamine and glutamate levels in nucleus accumbens of DBA/2J mice. Brain Res. 2008;1193:34–42.
33. Tarazi FI, Tomasini EC, Baldessarini RJ. Postnatal development of dopamine and serotonin transporters in rat caudate-putamen and nucleus accumbens septi. Neurosci Lett. 1998;254(1):21–4.
34. Teicher MH, Andersen SL, Hostetter Jr JC. Evidence for dopamine receptor pruning between adolescence and adulthood in striatum but not nucleus accumbens. Brain Res Dev Brain Res. 1995;89(2):167–72.
35. Chambers RA, Taylor JR, Potenza MN. Developmental neurocircuitry of motivation in adolescence: a critical period of addiction vulnerability. Am J Psychiatry. 2003;160(6):1041–52.
36. Wills TA, Vaccaro D, McNamara G. Novelty seeking, risk taking, and related constructs as predictors of adolescent substance use: an application of Cloninger's theory. J Subst Abuse. 1994;6(1):1–20.
37. Carrara-Nascimento PF, Griffin 3rd WC, Pastrello DM, et al. Changes in extracellular levels of glutamate in the nucleus accumbens after ethanol-induced behavioral sensitization in adolescent and adult mice. Alcohol. 2011;45(5):451–60.
38. Zhang Y, Loonam TM, Noailles PA, et al. Comparison of cocaine- and methamphetamine-evoked dopamine and glutamate overflow in somatodendritic and terminal field regions of the rat brain during acute, chronic, and early withdrawal conditions. Ann N Y Acad Sci. 2001;937:93–120.
39. Philpot R, Kirstein C. Developmental differences in the accumbal dopaminergic response to repeated ethanol exposure. Ann N Y Acad Sci. 2004;1021:422–6.
40. Philpot RM, Wecker L, Kirstein CL. Repeated ethanol exposure during adolescence alters the developmental trajectory of dopaminergic output from the nucleus accumbens septi. Int J Dev Neurosci. 2009;27(8):805–15.
41. Kalivas PW. Addiction as a pathology in prefrontal cortical regulation of corticostriatal habit circuitry. Neurotox Res. 2008;14(2–3):185–9.
42. Kalivas PW, Volkow ND. The neural basis of addiction: a pathology of motivation and choice. Am J Psychiatry. 2005;162(8):1403–13.
43. Garcia-Mijares M, Silva MTA. Dependência de drogas. Psicologia USP. 2006;17(4):27.
44. Schildkraut JJ, Schanberg SM, Breese GR, et al. Norepinephrine metabolism and drugs used in the affective disorders: a possible mechanism of action. Am J Psychiatry. 1967;124(5):600–8.
45. Zhang XL, Wang GB, Zhao LY, et al. Clonidine improved laboratory-measured decision-making performance in abstinent heroin addicts. PLoS One. 2012;7(1):e29084.
46. Seu E, Lang A, Rivera RJ, et al. Inhibition of the norepinephrine transporter improves behavioral flexibility in rats and monkeys. Psychopharmacol (Berl). 2009;202(1–3):505–19.
47. Dickson RA, Maki E, Gibbins C, et al. Time courses of improvement and symptom remission in children treated with atomoxetine for attention-deficit/hyperactivity disorder: analysis of Canadian open-label studies. Child Adolesc Psychiatry Ment Health. 2011;5:14.
48. Poschel BP, Ninteman FW. Norepinephrine: a possible excitatory neurohormone of the reward system. Life Sci. 1963;10:782–8.
49. Roberts DC, Corcoran ME, Fibiger HC. On the role of ascending catecholaminergic systems in intravenous self-administration of cocaine. Pharmacol Biochem Behav. 1977;6(6):615–20.

50. Jayanthi LD, Ramamoorthy S. Regulation of monoamine transporters: influence of psychostimulants and therapeutic antidepressants. AAPS J. 2005;7(3):E728–38.
51. Rothman RB, Baumann MH. Monoamine transporters and psychostimulant drugs. Eur J Pharmacol. 2003;479(1–3):23–40.
52. Rothman RB, Baumann MH, Dersch CM, et al. Amphetamine-type central nervous system stimulants release norepinephrine more potently than they release dopamine and serotonin. Synapse. 2001;39(1):32–41.
53. Sofuoglu M, Sewell RA. Norepinephrine and stimulant addiction. Addict Biol. 2009;14(2):119–29.
54. Emslie GJ. The psychopharmacology of adolescent depression. J Child Adolesc Psychopharmacol. 2012;22(1):2–4.
55. Laraia MT. Current approaches to the psychopharmacologic treatment of depression in children and adolescents. J Child Adolesc Psychiatr Nurs. 1996;9(1):15–26.
56. Sanders JD, Happe HK, Bylund DB, et al. Development of the norepinephrine transporter in the rat CNS. Neuroscience. 2005;130(1):107–17.
57. Sershen H, Shearman E, Fallon S, et al. The effects of acetaldehyde on nicotine-induced transmitter levels in young and adult brain areas. Brain Res Bull. 2009;79(6):458–62.
58. Krishnan V, Nestler EJ. The molecular neurobiology of depression. Nature. 2008;455(7215):894–902.
59. Monti JM, Jantos H. The roles of dopamine and serotonin, and of their receptors, in regulating sleep and waking. Prog Brain Res. 2008;172:625–46.
60. MacLean MR, Dempsie Y. Serotonin and pulmonary hypertension: from bench to bedside? Curr Opin Pharmacol. 2009;9(3):281–6.
61. Bellingham GA, Peng PW. Duloxetine: a review of its pharmacology and use in chronic pain management. Reg Anesth Pain Med. 2010;35(3):294–303.
62. Lam DD, Garfield AS, Marston OJ, et al. Brain serotonin system in the coordination of food intake and body weight. Pharmacol Biochem Behav. 2010;97(1):84–91.
63. Chan JS, Snoeren EM, Cuppen E, et al. The serotonin transporter plays an important role in male sexual behavior: a study in serotonin transporter knockout rats. J Sex Med. 2011;8(1):97–108.
64. Rapport MM, Green AA, Page IH. Serum vasoconstrictor, serotonin: isolation and characterization. J Biol Chem. 1948;176(3):1243–51.
65. Askenazy F, Caci H, Myquel M, et al. Relationship between impulsivity and platelet serotonin content in adolescents. Psychiatry Res. 2000;94(1):19–28.
66. Carver CS, Johnson SL, Joormann J. Two-mode models of self-regulation as a tool for conceptualizing effects of the serotonin system in normal behavior and diverse disorders. Curr Dir Psychol Sci. 2009;18(4):195–9.
67. Lambe EK, Krimer LS, Goldman-Rakic PS. Differential postnatal development of catecholamine and serotonin inputs to identified neurons in prefrontal cortex of rhesus monkey. J Neurosci. 2000;20(23):8780–87.
68. Andersen SL, Lyss PJ, Dumont NL, et al. Enduring neurochemical effects of early maternal separation on limbic structures. Ann N Y Acad Sci. 1999;877:756–9.
69. Schwandt ML, Lindell SG, Chen S, et al. Alcohol response and consumption in adolescent rhesus macaques: life history and genetic influences. Alcohol. 2010;44(1):67–80.
70. Van Oekelen D, Luyten WH, Leysen JE. 5-HT2A and 5-HT2C receptors and their atypical regulation properties. Life Sci. 2003;72(22):2429–49.
71. Morilak DA, Ciaranello RD. Ontogeny of 5-hydroxytryptamine2 receptor immunoreactivity in the developing rat brain. Neuroscience. 1993;55(3):869–80.
72. Carpenter-Hyland EP, Chandler LJ. Adaptive plasticity of NMDA receptors and dendritic spines: implications for enhanced vulnerability of the adolescent brain to alcohol addiction. Pharmacol Biochem Behav. 2007;86(2):200–8.
73. Doremus-Fitzwater TL, Varlinskaya EI, Spear LP. Motivational systems in adolescence: possible implications for age differences in substance abuse and other risk-taking behaviors. Brain Cogn. 2010;72(1):114–23.
74. Clark DB. Pharmacotherapy for adolescent alcohol use disorder. CNS Drugs. 2012;26(7):559–69.

André Luiz Monezi Andrade Graduated in Psychology, Positivo University. Msc and PhD in Psychobiology, Federal University of São Paulo/School of Medicine of São Paulo (UNIFESP/EPM—*Escola Paulista de Medicina*). Researcher in Psychobiology of drug dependence professor at the Department of Psychology from Anhembi Morumbi University—*Laureate International Universities*®.

Denise De Micheli BS in Psychology. PhD and postdoctoral fellowship in Science, Federal University of São Paulo (*Universidade Federal de São Paulo*—UNIFESP). Associate Professor and Chair of Medicine and Sociology of Drug Abuse, Department of Psychobiology, UNIFESP. Coordinator, Interdisciplinary Center for Studies in Neuroscience, Health and Education in Adolescence (*Centro Interdisciplinar de Estudos em Neurociência, Saúde e Educação na Adolescência*—Ciensea). Member of the Supervisory Board of the Research Incentive Fund Association (*Associação Fundo de Incentivo à Pesquisa*—AFIP). Member, Association for Medical Education and Research in Substance Abuse (Amersa). Collaborator, Institute for Translational Research in Adolescent Behavioral Health.

Eroy Aparecida da Silva BS in Psychology, São Paulo Organization of Education and Culture, Paulistana College. Specialist in Social and Occupational Psychology, Sedes Sapientiae Institute. Specialist in Family and Marriage Counseling, Pontifical Catholic University of São Paulo (*Pontifícia Universidade Católica de São Paulo*—PUC-SP). Doctor of Science, Department of Psychobiology, Federal University of São Paulo—UNIFESP. Full member, Brazilian Multidisciplinary Association of Studies on Drugs (*Associação Brasileira Multidisciplinar de Estudos sobre Drogas*—ABRAMD).

Maria Lucia Oliveira de Souza-Formigoni BS in Biological Sciences—biomedicine, Federal University of São Paulo/School of Medicine of São Paulo (UNIFESP/EPM). MS and PhD in Pharmacology, concentration in Psychopharmacology; Habilitation in Psychobiology, UNIFESP/EPM. Member, *Research Society on Alcoholism,* ISBRA (*International Society of Biomedical Research on Alcoholism*) and LASBRA (*Latin American Society of Biomedical Research on Alcoholism*) Assistant editor of the journal Addiction.

Francine de Oliveira Goeldner BS in Pharmacy, UFPR. Specialist and MS in Pharmacology, UFPR. PhD and postdoctoral fellowship in Psychobiology, Federal University of São Paulo (UNIFESP).

Part II
Adolescence, Drugs, and the Central Nervous System

Chapter 4
Neural Basis of Drug Addiction

Cassandra D. Gipson and Peter W. Kalivas

Drug addiction is defined as a chronically relapsing disorder, characterized by the compulsion to seek and take drugs, a loss of inhibition in the ability to control amount of intake, and the development of a negative hedonic state when access to drug is prohibited [1, 2]. According to Koob and Le Moal [1], there is a spiraling distress cycle of addiction that is comprised of social, psychiatric, and neurobiological mechanisms. The underlying neurobiological systems involved in the addiction cycle are quite complex, and no single mechanism mediates this process [3]. Although there is a constellation of underlying neural pathways and processes involved in the rewarding effects of drugs of abuse, the scope of this chapter will be limited to the mesocorticolimbic reward circuit.

The Reward System

Within the mesocorticolimbic reward circuit, key structures have been implicated in various stages of the addiction cycle (Fig. 4.1). Projections between these structures involve the transmission of both excitatory and inhibitory processes, and under conditions of chronic drug use, these processes are hypothesized to undergo allostatic changes which in turn lead to increased vulnerability to relapse [4]. Neuroplasticity and animal models of addiction will be discussed in further detail later in this chapter. Before delving into the specifics of the reward pathway involved in drug

C. D. Gipson (✉) · P. W. Kalivas
Department of Neuroscience, Medical University of South Carolina, Charleston, USA
e-mail: gipson@musc.edu

P. W. Kalivas
e-mail: kalivasp@musc.edu

D. De Micheli et al. (eds.), *Drug Abuse in Adolescence*,
DOI 10.1007/978-3-319-17795-3_4

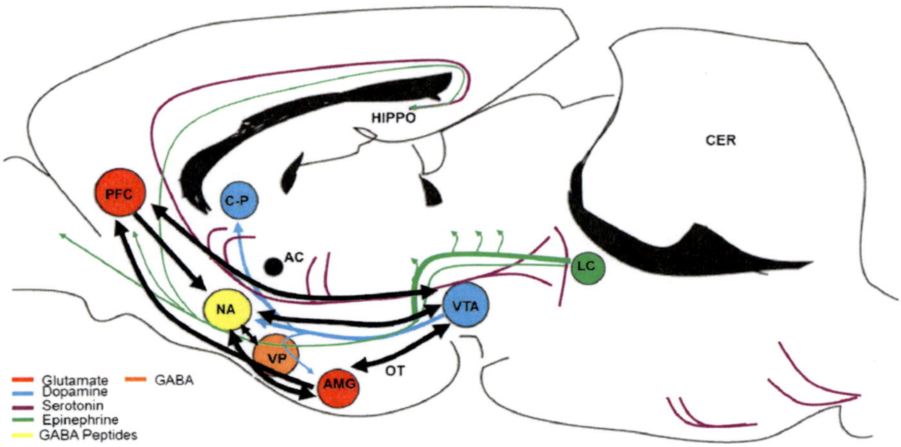

Fig. 4.1 Schematic of mesocorticolimbic brain circuitry involved in initiation and maintenance of addiction. *Black arrows* indicate projections between key mesocorticolimbic structures involved in addiction. (Modified from [3, 5]). *AC* anterior commissure, *AMG* amygdala, *CER* cerebellum, *C-P* caudate-putamen, *HIPPO* hippocampus, *LC* locus ceruleus, *OT* olfactory tubercle, *PFC* prefrontal cortex, *VP* ventral pallidum, *VTA* ventral tegmental area

addiction, it is first important to establish a foundation in biological function to understand the connections within this pathway.

Receptors

Biological function is accomplished through chemical signaling as a main mechanism of information transmission. Some drugs mimic a biological response (an agonist), and some prevent a biological response (an antagonist). They accomplish these effects by either competitively or noncompetitively binding to receptors and signaling a change in cell function, typically by changing ion conductance through the cell membrane or signaling through second messenger molecules to change intracellular functions, such as enzyme activity or gene expression. Agonists can be partial or full, such that they differ in the degree of biological response they illicit subsequent to their binding. A competitive antagonist will reversibly bind to a receptor at the endogenous ligand site. A noncompetitive antagonist, however, will bind to a site distinct from that of the endogenous ligand, and will exert their action via the other site. An uncompetitive antagonist requires receptor activation via an agonist prior to their binding to a separate, allosteric binding site.

In order for a ligand-binding molecule to be qualified as a receptor, it must satisfy certain criteria. The term *receptor* is reserved for molecules that exhibit both binding and signal generation functions. Additionally, receptors must display the following properties:

a. *Saturability*. Cell surface receptors engender the majority of receptors. There is a limited number of receptors per cell, thus a dose-response curve for ligand binding should occur. Specific receptor binding is characterized as high affinity of a ligand for a receptor, and low capacity, thus indicating saturability. Nonspecific binding, however, refers to high capacity and low affinity of a ligand for a receptor, thus this type of binding is nonsaturable.
b. *Specificity*. Binding only to the receptor of interest is termed specific binding. Nonspecific binding sites, however, engender the majority of receptors in tissue, thus specific binding is difficult to achieve. Binding assays in which displacement of the radiolabeled ligand of interest with various agonists/antagonists are necessary to determine specificity of a molecule to a receptor.
c. *Reversibility*. Transmitters, hormones, and most drugs bind in a reversible manner to receptors. In addition, a natural ligand of a reversible receptor should be recoverable in its non-metabolized form following reversible receptor binding.

Additionally, four distinct groups of receptors are known to be involved in signal transduction: (1) Ionotropic receptors, or ligand-gated channels, are located on the membrane and are composed of subunits and a central channel, which remains closed until activated following receptor binding. When the channel opens, Na^+, K^+, Ca^{2+}, or Cl^- flows into the cell, thus altering the membrane potential, either hyperpolarizing or depolarizing the cell depending on which ion is allowed passage. Ionotropic receptors allow for faster responses. (2) Metabotropic receptors, or G-protein coupled receptors, are the second and largest group of cell surface receptors and mediate slower responses. (3) Hormone receptors bind to lipophilic ligands, including steroid and thyroid hormones, vitamin D, and retinoic acids. These receptors have been found to be mostly intracellular (since hormones act on nuclear DNA to alter gene expression) but more recently steroid receptors have been found on membranes. (4) The fourth type of receptor, termed growth receptors, is the tyrosine kinase receptor located on the cell membrane. Activity of these receptors differs from other receptor subtypes in that kinase activity is part of the receptor. Additionally, activation of these receptors occurs by the presence of insulin and other growth factors, including nerve growth factor.

Neurotransmitters

Although there is a multitude of known neurotransmitters, as well as their detailed life cycles, the scope of this chapter will be limited to the function and location throughout the brain of amino acid neurotransmitters, acetylcholine, catecholamines, and serotonin.

a. *Amino Acid Neurotransmitters*. Amino acid neurotransmitters represent a large percentage of transmitters in the mammalian CNS, and are classified as either excitatory (glutamate, aspartate, cysteate, and homocysteate), which depolarize neurons, or inhibitory (γ-aminobutyric acid (GABA), glycine, taurine, and β-alanine), which hyperpolarize neurons. Although the dominant

Table 4.1 Identified neurotransmitter receptor subtypes. (Modified from [3])

Neurotransmitter	Receptor subtypes
Adrenergic	α_{1A}, α_{1B}, α_{1C}, α_{1D} α_{2A}, α_{2B}, α_{2C}, α_{2D} β_1, β_2, β_3
Cholinergic	Nicotinic: muscle, neuronal (α-bungarotoxin-insensitive), neuronal (α-bungarotoxin sensitive)
Muscarinic	M_1, M_2, M_3, M_4, M_5
Dopaminergic	D_1, D_2, D_3, D_4, D_5
GABAergic	$GABA_A$, $GABA_{B1a}$, $GABA_{B1\gamma}$, $GABA_{B2}$, $GABA_C$
Glutamatergic	NMDA, AMPA, kainate, $mGluR_1$, $mGluR_2$, $mGluR_3$, $mGluR_4$, $mGluR_5$, $mGluR_6$, $mGluR_7$
Histaminergic	H_1, H_2, H_3, H_4
Opioid	μ_1, μ_2, μ_3, δ_1, δ_2, κ_1, κ_2, κ_3
Serotoninergic	$5\text{-}HT_{1A}$, $5\text{-}HT_{1B}$, $5\text{-}HT_{1D}$, $5\text{-}HT_{1F}$, $5\text{-}HT_{2A}$, $5\text{-}HT_{2B}$, $5\text{-}HT_{2C}$, $5\text{-}HT_3$, $5\text{-}HT_4$, $5\text{-}HT_5$, $5\text{-}HT_6$, $5\text{-}HT_7$

neurotransmitter implicated in addiction has traditionally been dopamine (see below), recent evidence suggests that glutamate plays a large role in the transition from recreational drug use to chronic, escalating use characteristic of drug addiction [6]. Excitatory amino acid receptors include NMDA, AMPA, kainate, and metabotropic glutamate receptors (mGluRs; see Table 4.1). NMDA and AMPA receptors are distributed in parallel within the CNS, and they have been implicated in synaptic plasticity and the transition to addiction [6, 7]. Location of these receptors is widespread throughout the CNS, however, NMDA and AMPA receptors are largely found in hippocampus and cerebral cortex. In addition, mGluRs are also widely distributed throughout the CNS, and they are categorized into group I, II, or III mGluRs. Group I (including mGluR1 and mGluR5s) receptors are postsynaptic, and facilitate NMDA responses. These are involved in synaptic plasticity and long term potentiation (LTP) and long term depression (LTD). Group II (including mGluR2 and mGluR3s) receptors are located presynaptically, and are involved in inhibition of neurotransmitter release. Agonist stimulation of mGluR2/3 receptors has been shown to inhibit cocaine and heroin-seeking behavior in rats [8, 9]. Group III (including mGluR 4, mGluR7, and mGluR8s) are located on the presynaptic terminal, and are also involved in the inhibition of neurotransmitter release.

The major inhibitory neurotransmitter in the mammalian CNS is GABA. As such, it has been implicated in the pathogenesis of epilepsy. GABA receptors include $GABA_A$ and $GABA_B$ receptors. Ionotropic $GABA_A$ receptors are the most prevalent type of GABA receptor, and are associated with a chloride (Cl^-) channel. These receptors have been found to be involved in the sedative, anxiolytic, and anticonvulsant activity of clinically important drugs. The metabotropic $GABA_B$ receptor, however, acts as an autoreceptor in the regulation of GABA release, and is not as prevalent as $GABA_A$ receptors.

b. *Acetylcholine*. Acetylcholine (Ach) is the transmitter at neuromuscular junctions, and thus is expressed in both the peripheral nervous system (PNS) and the CNS. Cholinergic receptors are classified as with muscarinic or nicotinic (see Table 4.1), and are all slow, G-protein coupled receptors that either act directly on ion channels or are linked to second messenger systems. Depending on the cell type, these receptors either close potassium (K) channels, calcium (Ca^{2+}) channels, or Cl^- channels. In addition, nicotinic acetylcholine receptors (nAchRs) modulate the release of glutamate, dopamine, GABA, norepinephrine, and Ach, depending on location. Indeed, nAchRs have been implicated as potential therapeutic targets for Parkinson's disease, schizophrenia, attention deficit hyperactivity disorder, among many others. In addition, nAchRs are critical in nicotine addiction, and thus are a major therapeutic target in this disorder.
c. *Catecholamines*. Catecholamines are formed in the brain from an amino acid precursor, tyrosine hydroxylase, and include norepinephrine, epinephrine, and dopamine. Norepinephrine and epinephrine are metabolic products of dopamine. Norepinephrine is one of the most abundant neurotransmitters in the mammalian CNS, and plays an important role in selective attention, arousal, and stress [10]. As illustrated in Fig. 4.1, the noradrenergic system originates in the locus coeruleus (LC), and projects to the hippocampus (HIPPO), cerebellum (CER), and forebrain. According to Weinshenker and Shroeder [9], norepinephrine neurons from LC innervate the VTA and modulate excitatory burst firing of dopaminergic neurons. Thus, norepinephrine has been found to play a major role in preclinical models of drug addiction and relapse [11].

The neurotransmitter traditionally implicated in drug abuse is dopamine (DA; [6, 12]). DA is found in high concentrations within the VTA, and ultrashort, intermediate, and long projections exist in the distribution of this neurotransmitter throughout the mammalian CNS. Long projections link VTA and substantia nigra cells with the neostriatum (including caudate and putamen), limbic cortex (including prefrontal cortex (PFC), cingulate, and entorhinal areas), and limbic structures (including septum, olfactory tubercle (OT), nucleus accumbens (NA), septi, AMG, and piriform cortex (see Fig. 4.1 for a simplified diagram of these projections). The long projections are classified as mesocortical and mesolimbic DA projections. DA release within the PFC-NA-VP series circuit has been found to mediate cocaine-seeking behavior in a preclinical model of drug relapse [13]. In addition, imaging studies with drug addicts (including nicotine, cocaine, alcohol, and heroin-addicted subjects) have shown decreased levels of DA $D_{2/3}$ receptors [14]. DA has been implicated in "wanting" of hedonic rewards, and for enhancing incentive salience [15].

d. *Serotonin*. Serotonin (5-HT) has been implicated in various types of mental illness, including depression. 5-HT is synthesized from tryptophan, which is derived primarily from the diet. 5-HT pathways in the mammalian CNS are restricted to clusters of cells located near the raphe regions of the pons and upper brain stem (see Fig. 4.1). In addition, 5-HT receptors are located both in the PNS and CNS. 5-HT within medial PFC and OFC has been implicated in impulsive choice [16]. Specifically, deficiencies in 5-HT have been implicated in a disruption in normal inhibitory processes involved in motivation and decision making [17]. Indeed, decreased 5-HT activity has been associated with impulsivity and maladaptive decision making, such as violence, suicide, and pathological gambling behaviors [18, 19].

The Mesocorticolimbic Reward Circuit

The model of the mesocorticolimbic reward circuit illustrated in Fig. 4.1 is the result of decades of clinical and preclinical research, including data from imaging techniques in humans with addictive behaviors, behavioral pharmacology, and cellular physiology and biology in nonhuman animals. Key nodes in the mesocorticolimbic reward circuit include PFC, AMG, VP, VTA, and striatum (including NAshell and core, caudate and putamen). Dopaminergic and glutamatergic connections between these key structures have been implicated in the transition from recreational to escalated drug use characteristic of drug addiction [7, 20]. For example, DA transmission from PFC to NA has been found to play an important role in reinstatement of cocaine-seeking behavior in rats [13]. In the extended access model of drug intake, steady levels of DA were elevated in NA of rats given daily long access of cocaine self-administration compared to rats given daily short access to cocaine self-administration, and this elevation was directly related to the amount of daily cocaine intake [21]. In addition, Kalivas et al. [20] postulate that changes in glutamate in NA may play an important role in relapse.

Animal Models of Drug Addiction

To understand the advances made in the field of addiction, one must first become acquainted with animal models of drug addiction. Animal models are imperative in furthering our understanding of the biology and pathophysiology of substance abuse [22]. Although various operant and Pavlovian behavioral models are employed in preclinical drug addiction research, this chapter will focus on operant drug self-administration paradigms designed to examine the various stages of the addiction cycle.

Historically, animal behavior was studied to help us better understand human behavior. According to Overmier [23], "Models are basic and powerful tools in science," and they allow for more controlled, simple, and less expensive conditions [24]. Many animal models of drug addiction are based on the analysis of behavioral output during various schedules of reinforcement, which were initially developed by Ferster and Skinner [25]. Primarily, animal models of drug addiction include self-administration, in which animals are placed in operant chambers, and completion of a schedule of reinforcement via lever presses are accompanied with intravenous drug delivery (Fig. 4.2). Typically, fixed-ratio (FR) schedules are employed,

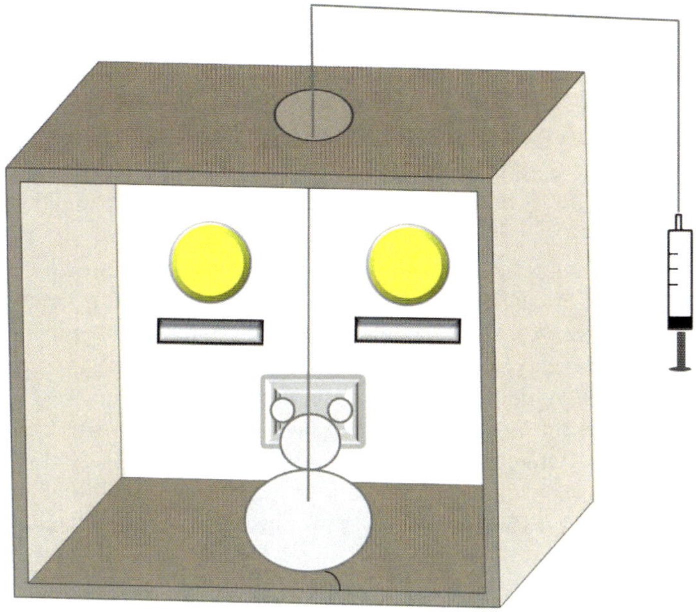

Fig. 4.2 Illustration of an operant conditioning chamber used in drug self-administration. Lights, levers, and a food receptacle are located on an intelligence panel, and an infusion pump is located outside the chamber. Responses to the active lever typically results in the delivery of one intravenous drug infusion

such that an animal must press a lever a fixed number of times prior to drug reinforcement delivery. Additionally, progressive-ratio (PR) schedules are used to examine the reinforcing efficacy of a drug (i.e., the likelihood that a drug will serve as a reinforcer). In this higher-order schedule, an animal is required to emit an increasing number of responses for each successive reinforcer. Self-administration is frequently used in animal models of addiction, as it more closely models the human condition than experimenter-delivered drug.

To study the various stages of addiction, behavioral paradigms have been developed using the self-administration model. Stages of acquisition, maintenance, escalation, withdrawal, and relapse have been examined in this way.

a. *Acquisition.* To examine the acquisition stage of addiction, autoshaping procedures have been developed such that an animal is trained in the contingencies associated with an active (lever presses to this lever will result in the presentation of food or drug delivery) and inactive lever (lever presses to this lever will result in no programmed consequence). Carroll and Lac [26] developed a drug autoshaping procedure in which the active lever is extended on a fixed time schedule (the lever will extend every 60 s), and a lever press will result in the delivery of a drug infusion, along with the illumination of a cue light above the lever (see Fig. 4.2). If no lever press is emitted within 15 s, a drug infusion + cue light will occur non-contingently to promote acquisition of a Pavlovian association between the cue light (a conditioned stimulus, or CS), and the drug infusion (an unconditioned stimulus, or US). Differences in acquisition of drug self-administration can be measured.

b. *Maintenance*. Using fixed-ratio schedules of reinforcement, animals are placed into operant chambers for limited access periods (e.g., 1 or 2 h of drug self-administration per day). Amount of intake can be measured and manipulated via injections of certain compounds designed to increase or decrease drug intake. Additionally, dose-response curves can be generated using limited access procedures.
c. *Escalation*. An animal model has been developed to examine adaptations that occur during the escalation phase of addiction [27–30]. In this behavioral paradigm, animals are exposed to extended access sessions (e.g., 6 h of drug self-administration per day), and drug intake is measured. With stimulants such as cocaine (e.g., [29]), d-amphetamine [31], and methamphetamine [32], as well as the opioid heroin [33], drug intake generally increases across sessions with extended access, whereas limited access groups show relatively stable levels of intake across sessions. Although escalated intake may involve other processes such as learning and stimulus control [34], the extended access model has been used to examine possible neurobiological adaptations that only occur under conditions of increased intake (e.g., increased intracranial self-stimulation thresholds following escalated intake in extended access sessions; [35]).
d. *Withdrawal*. Animal models of withdrawal include either extinction training, or withdrawal without extinction training. In a typical behavioral paradigm, animals are trained to self-administer drug (e.g., cocaine). Following response stability, animals are either given extinction training in which the levers are extended during daily sessions, but responses to either lever result in no programmed consequence, or animals are placed back into their home cages for a specified period of time prior to reinstatement testing (termed abstinence). Although withdrawal with or without extinction can be used, recent research has found that extinction training may be necessary to activate the reward circuitry involved in addiction [36].
e. *Relapse*. A preclinical model used to measure relapse is the reinstatement behavioral paradigm (see Fig. 4.3 for a schematic of reinstatement protocol). In a

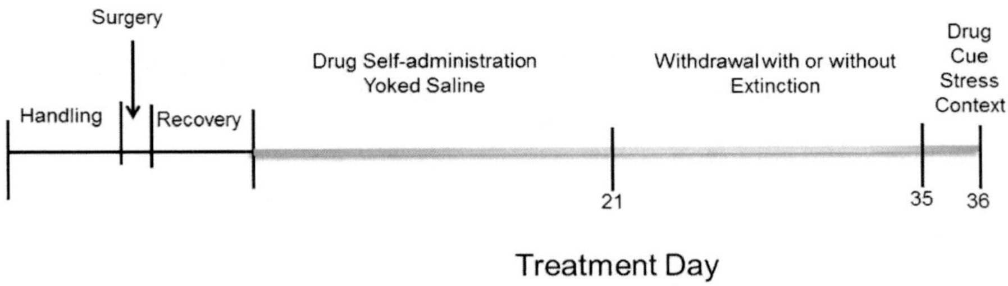

Fig. 4.3 Schematic of a typical reinstatement paradigm. Animals are implanted with jugular catheters, and are then trained to intravenously self-administer drug. Following response stability, animals are placed into withdrawal with or without extinction training. Reinstatement priming stimuli are then presented and drug-seeking behavior (lever press responses) is measured

typical reinstatement paradigm, animals are trained to self-administer drug (e.g., cocaine or heroin). To ensure acquisition of self-administration, rats must meet a set of response stability criteria prior to withdrawal. Following response stability, animals are placed into withdrawal with or without extinction. Following response stability in extinction or a specified period of time (without extinction training), animals are given a reinstatement priming stimulus. There are several priming stimuli used in the reinstatement model, including (1) a priming injection of the previously self-administered drug; (2) a cue previously associated with the delivery of a drug infusion; (3) a pharmacological or physical stressor, such as yohimbine or foot shock, respectively; or (4) placement back into the context in which the animal learned to self-administer drug. In the fourth paradigm, animals are trained in one context (context A), extinguished to the cues associated with drug in another distinct context (context B), and subsequently placed back into context A during reinstatement testing (termed the ABA renewal paradigm; [37–39]).

Reinstatement is a frequently used behavioral assay to measure drug-seeking following abstinence when a drug priming injection, stress, or environmental cues evoke this behavior. There is considerable debate, however, over the construct validity of this model [40]. Construct validity involves the ability of a model to illustrate similarity in the mechanisms underlying behavior between the modeled behavior and the behavior in the model. Although this model appears to have predictive validity, there are challenges in successfully achieving construct validity. When assessing the construct validity of reinstatement, there are many issues to consider. First, the drug-free state occurs under different conditions in the model than in the human condition. Specifically, rats extinguish lever pressing in self-administration because drug is no longer available, whereas in humans, attempting to quit drug use results from a complex set of choices in which the negative consequences of continuing to use drugs outweigh the reinforcing ones. Secondly, human relapse does not usually result from the types of drug priming and cue-induced reinstatement contingencies employed in the model. Specifically, these are given non-contingently in the animal model, whereas in the human condition, relapse typically occurs following contingent exposure to drugs or non-contingent exposure to drug-related cues. Third, many stress-induced reinstatement models employ footshock, which is not a typical occurrence encountered by humans. Lastly, risk of relapse appears to decrease with extended abstinence in humans, whereas the magnitude of reinstatement does not decrease over time in the preclinical model.

The Role of Learning and Memory Processes in Addiction

Drug addiction is characterized by transitions from initiation, to maintenance, to escalation. This process is viewed as the transition from voluntary, goal-directed action to habitual, uncontrolled behavior, and is dependent upon interactions between classical conditioning and instrumental conditioning learning processes [41, 42].

Additionally, research has implicated normal mechanisms of reward learning and memory processes in addiction [43, 44].

a. *Classical Conditioning*. A basic form of learning is classical conditioning (also known as Pavlovian conditioning). This learning process involves either excitatory or inhibitory associations between two stimuli. In either case, a CS, or a previously neutral stimulus that does not elicit a response, gains predictive value of the occurrence of a US with training. Animals tend to approach and interact with stimuli that signal the delivery of food, thus sign tracking has been investigated using discrete stimuli presented immediately before the delivery of food (the first experiment was reported by Brown & Jenkins [45]). More recently, research examining learning and incentive salience has found that a CS will elicit individual differences in CRs, such that some animals are designated as sign trackers, and others are goal trackers [46, 47] Sign trackers tend to approach the discrete stimulus associated with the delivery of food (e.g., the lever), whereas goal trackers tend to approach the food apparatus (e.g., the food receptacle). Individual differences in CRs have been found to predict novelty seeking behavior and acquisition of cocaine self-administration, such that goal tracking rats exhibit increased novelty-seeking behavior and acquisition of cocaine self-administration [46].
b. *Instrumental Conditioning*. Instrumental behavior occurs because it was previously involved in producing certain consequences [24]. Modern approaches to studying instrumental conditioning in drug addiction include operant responses (such as a lever press) which lead to the delivery of an appetitive stimulus (such as an intravenous drug infusion). As mentioned previously, this procedure of positive reinforcement is termed self-administration (see Fig. 4.2 for an illustration of an operant conditioning chamber). According to Everitt and Robbins [41], drug self-administration initially involves action-outcome learning (goal-directed behavior), and later transitions to stimulus-response associations (habit formation) which maintains drug seeking behavior. In the next section, the underlying neurobiological changes associated with this transition are discussed.
C. *Memory Processes*. According to Volkow et al. [44], learning and memory processes centered in the amygdala and hippocampus brain structures contribute largely to drug relapse. In this hypothesis, conditioned cues (e.g., a cigarette label) trigger memories of the properties of a drug, and in turn produce drug-seeking behavior. Indeed, the presentation of cues previously associated with drug delivery can increase dopamine and promote reinstatement of drug-seeking behavior in rats [48].

The hippocampus is a structure implicated in associative memory, thus it encodes and consolidates information from the environment, and plays a role in learning the relationships between stimuli and the environment. In addition, dopaminergic and glutamatergic pathways in cortex, limbic system, and basal ganglia have been implicated in motivation, learning, and memory [49]. Thus, neural plasticity found to underlie the progression of drug addiction shares commonalities with the neural mechanisms underlying learning of natural reward value and memory [50].

The Neural Stages of Addiction

The neural mechanisms underlying various stages of addiction have been extensively examined at both clinical and preclinical levels. In this section, mechanisms underlying initiation (including possible risk factors involved in the initiation of drug use), escalation (also discussed is the transition from goal-directed behaviors to compulsive, habit formation), withdrawal, and relapse are discussed.

a. *Initiation*. Individual differences exist in risk vulnerability to drug abuse. According to de Wit [51], impulsivity can act as both a determinant and a consequence of drug use. In addition, impulsivity acts as a risk factor during initial recreational use of drugs, as well as during escalation of drug use and during relapse in the downward spiral of addiction [52]. Clinical and preclinical research on the neurobiological roots of impulsivity has implicated key structures in the mesocorticolimbic reward pathway, including PFC, orbitofrontal cortex (OFC), AMG, and NA.

PFC has been implicated in preclinical models of impulsive choice (choice between a smaller, more immediate reward, and a larger, more delayed reward). Cardinal et al. [53] found that rats increased choice of the smaller, more immediate reward when delays were short, but decreased choice of this option when delays were relatively long at the end of the session during a progressive delay task (in which delay to the larger, delayed reward was increased from 0 to 60 s across a session) following lesions to the medial wall of PFC (including both prelimbic and infralimbic cortices). Although these data do not conclusively confirm a change in impulsive choice in this study, they do implicate this region in decision making processes, although the exact role is still unclear [54]

OFC has been found to play an important role in impulsivity (both delay aversion and impulsive action; [55]). In cocaine-dependent subjects, reduced OFC metabolic activity was associated with reduced striatal D_2 receptor availability [56]. Additionally, OFC volume is markedly decreased in cocaine-dependent individuals [57]. In preclinical research, lesions to OFC have been found to both increase [58] and decrease [59] delay aversion (i.e., intolerance to delay of gratification), depending on the behavioral task. The conflicting results found in these two reports highlight an important issue in the impulsivity literature. Under the umbrella term "impulsivity," both personality scales and behavioral tasks are included. Within each of these types of measures, there are numerous variations of scales and tasks that attempt to measure the same construct, except nomenclature varies from task to task. Thus, the breadth of measures has led to difficulty in interpretation of results [60].

The NA, a key node in the mesocorticolimbic reward pathway, has been implicated in reward-related behavior. Within the NA, the core and shell have been shown to play distinct and critical roles in impulsivity. The NA core is involved in acquisition of instrumental responses, as lesions to this area inhibit performance of autoshaped responses [61] and disrupt Pavlovian-instrumental transfer, which is the facilitation of instrumental responses by the presentation of a CS [62]. In addition, lesions to the NA (both core and shell) decreased sensitivity to changes in delay to reinforcement in an adjusting delay discounting task [63].

Another key region in the mesocorticolimbic reward pathway, the AMG, has been implicated in impulsivity. Specifically, the basolateral amygdala (BLA) which shares many reciprocal connections with OFC, has been implicated in goal-directed behavior and impulsive choice. Lesions to the BLA increase choice of the smaller, more immediate reward in delay discounting [64].

b. *Escalation.* The phase of addiction associated with neurobiological changes occurring from chronic, compulsive drug use has been termed the escalation phase, as well as the phase in which behavior changes from controlled and goal-directed, to uncontrolled habit-formation associated with compulsion ([41, 65], respectively). Both of these hypotheses center on a fundamental change that occurs as a result of chronic drug use. In the escalation hypothesis of drug addiction, this change is termed "allostasis," in which the brain reestablishes stability following chronic drug use [65]. According to Everitt and Robbins [41], the change from voluntary, goal-directed drug use to uncontrolled, compulsive drug use is the result of a fundamental change at the neural level in control from the prefrontal cortex to striatum. Additionally, Kalivas [7] postulates that the NA serves as a gateway in the transition from limbic to motor control in addiction. According to Kalivas and Volkow [6], neuron function changes as a result of the transition from recreational to escalated drug use. In this transition, it has been postulated that there is a switch from dopaminergically-stimulated to glutamatergically-stimulated behaviors.
c. *Withdrawal.* According to Koob et al. [41], changes in the neurobiological systems involved in the development of drug withdrawal may underlie the changes from drug use to drug addiction. In this hypothesis, the changes that occur during withdrawal from drugs include opposite compensatory mechanisms from what occurred during drug use (the opponent-process theory of motivation initially proposed by [66]). Although dependence and withdrawal were previously known as hallmarks of addiction, it is now recognized that these symptoms are neither necessary nor sufficient for addiction without compulsion [43, 67]. Although this may be true, Winstanley et al. [68] postulate that withdrawal may lead to increased impulsivity, and thus increased vulnerability to relapse.

At the preclinical level of analysis, the duration of withdrawal has a profound effect on reinstatement of drug seeking behavior induced by cues previously associated with drug and stress (also known as the incubation of drug craving effect; [69, 70]). Additionally, Knackstedt et al. [36] found that extinction training, rather than withdrawal without extinction, was necessary to engage the prelimbic-NAcore circuit in cocaine seeking behavior, as inhibition of the prelimbic cortex did not inhibit reinstatement when animals were placed back into the environment previously associated with cocaine following forced abstinence.

d. *Relapse.* The role of the limbic-striato-pallidal neural circuitry in relapse has been examined [13]. Specifically, projections between key regions such as PFC, VTA, nucleus accumbens core (NAcore), and VP have been implicated in cocaine-induced reinstatement (see Fig. 4.1). Two subcircuits have been identified as being either limbic (ventral PFC, nucleus accumbens shell (NAshell), medial VP, amygdala, and VTA) or motor (dPFC, NAcore, dorsolateral VP, and substantia nigra; [13, 71]), the former of which has been traditionally associated with craving and relapse [72, 73], and the latter of which has been postulated to be involved in the compulsive or automatic behaviors involved in relapse [13].

The Reward System in the Adolescent Brain

The adolescent brain is dynamic in its neurochemistry, fiber construction, and tissue composition. Additionally, this is a period of development for prefrontal cortex, limbic structures, and white matter pathways [74, 75]. Although these structures develop in order to assist with navigation in a complex environment throughout adulthood, adolescence remains a time of heightened risk taking and sensation seeking, thus rendering individuals more vulnerable to initiate use of alcohol and drugs [76]. Treatment of drug abuse in adolescents has proven difficult, as many youths resume taking drugs following cessation of treatment [77]. Indeed, adolescence is a period of heightened impulsivity, and impulsivity has been linked to various stages of drug use [51, 68].

Although impulsivity has been linked to increased risk of substance abuse, it likely has its roots in evolution. There is an adolescent spike in the urgency (or mood-based rash action) facet of impulsivity as measured on a personality scale (the Urgency, lack of Perseverance, lack of Planning, and Sensation Seeking scale; the UPPS), and this may have evolutionarily-driven adaptive functions [78]. Adolescence-limited increases in emotionality and risk taking have been found to occur across species [76]. This may be adaptive for helping adolescents to engage in the next steps in development, which is leaving the natal home. Adolescent spikes in other facets of impulsivity, such as sensation seeking, may also be adaptive for procreation, as these spikes may enable adolescents to take risks for the survival of the species. A "normal" sort of impulsivity may be adaptive for finding new food sources when food is scarce, as it entails taking risks in eating new foods that may be harmful. According to Dickman [79], impulsivity can be broken down into two types, both functional and dysfunctional. These both lead to fast and error-prone action, but are quite different in meaning. Functional impulsivity refers to the willingness and ability to take risks in appropriate and necessary situations in which this behavior results in a benefit for survival. Dysfunctional impulsivity, however, refers to the tendency to engage in thoughtlessness and the inability to plan ahead and anticipate negative consequences. Thus, although increased impulsivity and risk taking during adolescence can lead to increased maladaptive behaviors, the neurobiological changes during adolescence that underlie increases in risk taking may serve an important evolutionary function.

a. *Mesocorticolimbic Reward Circuitry in Adolescence.* Adolescence-induced neural changes in mesocorticolimbic reward circuitry include alterations in regions such as PFC, AMG, and NA. Interestingly, these regions have also been implicated in various forms of impulsivity, and adolescence is associated with increased impulsivity, vulnerability to drug addiction, and novelty seeking as a result of developmental changes in frontal cortical and subcortical monoaminergic systems [17]. In PFC, cholinergic innervation and DA transporter (DAT) density increases during adolescence [80, 81]. Further evidence that PFC undergoes extensive changes during adolescence, dendritic spines in monkey prefrontal cortex undergo substantial pruning during adolescence [82], thus furthering the notion that this key structure in the mesocorticolimbic reward circuitry undergoes extensive restructuring and topographical modification. Additionally, connections between PFC and NA continue to develop throughout adolescence, with increases in the number of projecting pyramidal cells to NA, as well as an increase in the proportion of these cells that express DA D_1 receptors [83]. Importantly, drug exposure during the adolescent density peak of DA D_1 receptors may lead to neurobiological alterations in PFC such that it becomes programmed to be more vulnerable to drug addiction by forming strong associations with drugs that are resistant to extinction, and thus reinstatement occurs at a higher rate in adolescents [83, 84]. According to Kalivas et al. [20], the dopamine projection from PFC to NA is involved in reinstatement of cocaine seeking in rats, thus the findings from Brenhouse et al. [83] suggest that adolescence is a significant time point in the development of a critical projection involved in relapse. In addition, clinical research on addicts has found increased blood flow and metabolism in these two brain regions following the presentation of drug-associated cues using neuroimaging techniques, and that this increased activity is associated with increased self-reports of "drug craving" [85].

Prefrontal cortical inputs to basal AMG also undergo extensive restructuring during adolescence. Specifically, using tract-tracing and gene expression profiling methods, Cressman et al. [86] found changes in the medial PFC-basal AMG projection across early adolescence (postnatal day [PND] 25), late adolescence (PND 45), and into adulthood (PND 90) such that extensive pruning of neurons and axons originating in medial PFC and terminating in basal AMG occurred, with a 50% decrease from PND 45 to PND 90.

b. *Motivational Alterations in Adolescence*. As a result of the brain transformations in adolescence, there are alterations in motivational and reward systems, such as changes in incentive salience, learning, and value of reward-related stimuli [15, 87].

Interestingly, although adolescence is a period of development marked by increased impulsivity and drug abuse vulnerability, preclinical models of relapse have illustrated a potential resistance in adolescent rats (PND 28-60; [76]) compared to adult rats in incubation of drug craving during withdrawal, such that adolescent rats showed less cue-induced reinstatement of cocaine-seeking behavior compared to adult rats following 30 days of abstinence [88]. In addition, adolescent rats have also shown elevated cocaine seeking behavior during maintenance of cocaine self-administration, as well as during extinction and during cocaine- and stress-induced reinstatement compared to adult rats [89]. It has been postulated that adolescent rats are more sensitive to the reinforcing effects of cocaine and show decreased ability to inhibit previously reinforced behavior in extinction [89].

c. *Risky Decision Making and Social Interactions in Adolescence*. Adolescence is a time of heightened social interactions, and thus peer influence plays an important role in risky decision making and initiation of drug use during this critical developmental time period [90, 91]. In a study examining risk taking in adolescent versus adult human peer groups, Gardner and Steinberg [91] found that adolescents are more likely to take risks and exhibit risky decision making in a driving task when in the presence of peers than when alone compared to their adult counterparts, thus supporting the notion that peer influence plays a role in adolescent risky behavior. In addition, risky decision making decreased with age. Although recent preclinical work has shown increased drug intake with exposure to novel conspecifics (e.g., social facilitation of d-amphetamine self-administration; [92]), developing preclinical models of adolescent social facilitation of drug self-administration is challenging due to the short period of adolescence, as well as the extended training necessary to achieve stable rates of drug self-administration in rodents.

Similar to human adolescents, social interactions have been shown to be rewarding in adolescent rats [93]. According to Varlinskaya and Spear [94], early adolescent rats exhibit higher levels of ethanol-induced facilitation of social behavior, and are less sensitive to the social suppressive effects of high doses of ethanol compared to their mid- or older adolescent counterparts. Thus, the unique pattern of sensitivity to the effects of ethanol on social behavior in early adolescence may lead to higher risk for development of substantial alcohol use in this age group.

Summary

The future of addiction research lies in improving the translational validity of preclinical models, as well as integrating studies at multiple levels of analysis (e.g., behavior, brain imaging, neuroanatomy, circuitry, genetic factors, and cellular and molecular biology). Additionally, improvement of preclinical models of adolescent substance use is imperative in understanding the early neurobiological risk factors of addiction. For example, one issue in achieving this goal lies in the difficulty of implementing self-administration in adolescent animals due to necessary extended training and the short window of adolescence in rodents. Thus, improvement of preclinical models of adolescent substance use is necessary to further understand the neurobiological underpinnings of addiction during this critical period of development, as well as the possible increased risk for addiction in adulthood with exposure to drugs of abuse during adolescence.

References

1. Koob GF, Le Moal M. Drug abuse: hedonic homeostatic dysregulation. Science. 1997;278: 52–8.
2. Koob GF, Le Moal M. Drug addiction, dysregulation of reward, and allostasis. Neuropsychopharmacology. 2001;24(2):97–126.
3. Iversen LL, Iversen SD, Bloom FE, Roth RH. Introduction to neuropsychopharmacology. In: Introduction to neuropsychopharmacology. New York: Oxford University Press; 2009.
4. Le Moal M, Koob GF. Drug addiction: pathways to the disease and pathophysiological perspectives. Eur Neuropsychopharmacol. 2007;17:377–393.
5. Kalivas PW. Glutamate systems in cocaine addiction. Curr Opin Pharmacol. 2004;4(23): 23–9.
6. Kalivas PW, Volkow ND. The neural basis of addiction: a pathology of motivation and choice. Am J Psychiatry. 2005;162:1403–13.
7. Kalivas PW. The glutamate homeostasis hypothesis of addiction. Nat Rev Neurosci. 2009;10:561–72.
8. Bossert JM, Gray S, Lu L, Shaham Y. Activation of group II metabotropic receptors in the nucleus accumbens shell attenuates context-induced relapse to heroin seeking. Neuropsychopharmacology. 2006;31:2197–209.
9. Peters J, Kalivas PW. The group II metabotropic glutamate receptor agonist, LY379268, inhibits both cocaine- and food-seeking behavior in rats. Psychopharmacology (Berl). 2006;186:143–9.
10. Weinshanker D, Schroeder JP. There and back again: a tale of norepinephrine in drug addiction. Neuropsychopharmacology. 2007;32:1433–51.
11. Aston-Jones G, Harris GC. Brain substrates for increased drug seeking during protracted withdrawal. Neuropsychopharmacology. 2004;47:167–79.
12. Volkow ND, Fowler JS, Wang GJ, Baler R, Telang F. Imaging dopamine's role in drug abuse and addiction. Neuropharmacology. 2009;56:3–8.
13. McFarland K, Kalivas PW. The circuitry mediating cocaine-induced reinstatement of drug-seeking behavior. J Neurosci. 2001;21(21):8655–63.
14. Cosgrove K. Imaging receptor changes in human drug abusers. Curr Top Behav Neurosci. 2010;3:199–217.
15. Berridge KC. The debate over dopamine's role in reward: the case for incentive salience. Psychopharmacology. 2007;191:391–431.
16. Winstanley CA, Theobald DEH, Dalley JW, Cardinal RN, Robbins TW. Double dissociation between serotonergic and dopaminergic modulation of medial prefrontal and orbitofrontal cortex during a test of impulsive choice. Cereb Cortex. 2006;16(1):106–14.
17. Chambers RA, Taylor JR, Potenza MN. Developmental neurocircuitry of motivation in adolescence: a critical period of addiction vulnerability. Am J Psychiatry. 2003;160:1041–52.
18. Brown GL, Linnoila MI. CSF serotonin metabolite (5-HIAA) studies in depression, impulsivity, and violence. J Clin Psychiatry. 1990;51:31–41.
19. Nordin C, Eklundh T. Altered CSF 5-HIAA disposition in pathological male gamblers. CNS Spectr. 1999;4:25–33.
20. Kalivas PW, Volkow N, Seamans J. Unmanageable motivation in addiction: a pathology in prefrontal-accumbens glutamate transmission. Neuron. 2005;45:647–50.
21. Ahmed SH, Lin D, Koob GF, Parsons LH. Escalation of cocaine self administration does not depend on altered cocaine-induced nucleus accumbens dopamine levels. J Neurochem. 2003;86(1):102–13.
22. Lynch WJ, Nicholson KL, Dance ME, Morgan RW, Foley PL. Animal models of substance abuse and addiction: implications for science, animal welfare, and society. Comp Med. 2010;60(3):177–88.
23. Overmier JB. On the nature of animal models of human behavioral dysfunction. In: Haug M, Whalen RE, Editors. Animal models of human emotion and cognition. Washington, DC: American Psychological Association; 1999. pp. 15–24.

24. Domjan M. The principles of learning and behavior. 5 Ed. Belmont: Wadsworth; 2003.
25. Ferster CB, Skinner BF. Schedules of reinforcement. New York: Appleton Century-Crofts; 1957.
26. Carroll ME, Lac ST. Autoshaping i.v. cocaine self-administration in rats: effects of nondrug alternative reinforcers on acquisition. Psychopharmacology (Berl). 1993;110:5-12.
27. Ahmed SH, Koob GF. Transition from moderate to excessive drug intake: change in hedonic set point. Science. 1998;282:298–300.
28. Ahmed SH, Koob GF. Long-lasting increase in the set point for cocaine self-administration after escalation in rats. Psychopharmacology. 1999;146:303–12.
29. Ahmed SH, Koob GF. Changes in response to a dopamine receptor antagonist in rats with escalating cocaine intake. Psychopharmacology. 2004;172:450–4.
30. Ahmed SH, Koob GF. Transition to drug addiction: a negative reinforcement model based on an allostatic decrease in reward function. Psychopharmacology. 2005;180:473–90.
31. Gipson CD, Bardo MT. Extended access to amphetamine self administration increases impulsive choice in a delay discounting task in rats. Psychopharmacology. 2009;207:391–400.
32. Kitamura O, Wee S, Specio SE, Koob GF, Pulvirenti L. Escalation of methamphetamine self administration in rats: a dose effect function. Psychopharmacology. 2006;186(1):48–53.
33. Ahmed SH, Walker JR, Koob GF. Persistent increase in the motivation to take heroin in rats with a history of drug escalation. Neuropsychopharmacology. 2000;22:413–21.
34. Beckmann JS, Gipson CD, Marusich JA, Bardo MT. Stimulus control of escalated cocaine intake and escalation under short access sessions. (in review)
35. Ahmed SH, Kenny PJ, Koob GF, Markou A. Neurobiological evidence for hedonic allostasis associated with escalating cocaine use. Nature Neurosci. 2002;5:625–6.
36. Knackstedt LA, Moussawi K, Lalumiere R, Schwendt M, Klugmann M, Kalivas PW. Extinction training after cocaine self-administration induces glutamatergic plasticity to inhibit cocaine seeking. J Neurosci. 2010;30(23):7984–92.
37. Bouton ME, Bolles RC. Contextual control of the extinction of conditioned fear. Learn Motivation. 1979;10:445–66.
38. Crombag HS, Shaham Y. Renewal of drug seeking by contextual cues after prolonged extinction in rats. Behav Neurosci. 2002;116:169–73.
39. Crombag HS, Bossert JM, Koya E, Shaham Y. Context-induced relapse to drug seeking: a review. Philos Trans Royal Soc. 2008;363:3233–43.
40. Epstein DH, Preston KL, Stewart J, Shaham Y. Toward a model of drug relapse: an assessment of the validity of the reinstatement procedure. Psychopharmacology (Berl). 2006;189(1): 1–16.
41. Everitt BJ, Robbins TW. Neural systems of reinforcement for drug addiction: from actions to habits to compulsion. Nat Neurosci. 2005;8(11):1481–9.
42. Zapata A, Minney VL, Shippenberg TS. Shift from goal-directed to habitual cocaine seeking after prolonged experience in rats. J Neurosci. 2010;30(46):15457–63.
43. Hyman SE, Malenka RC, Nestler EJ. Neural mechanisms of addiction: the role of reward-related learning and memory. Ann Rev Neurosci. 2006;29:565–98.
44. Volkow ND, Wang GJ, Fowler JS, Tomasi D, Telang F, Baler R. Addiction: decreased reward sensitivity and increased expectation sensitivity conspire to overwhelm the brain's control circuit. BioEssays. 2010;32(9):748–55.
45. Brown PL, Jenkins HM. Auto-shaping the pigeon's key peck. J Exp Anal Behav. 1968;11: 1–8.
46. Beckmann JS, Marusich JA, Gipson CD, Bardo MT. Novelty seeking, incentive salience and acquisition of cocaine self-administration in the rat. Behav Brain Res. 2010;216(1):159–65.
47. Robinson TE, Flagel SB. Dissociating the predictive and incentive motivational properties of reward-related cues through the study of individual differences. Biol Psychiatry. 2009;65:869–873.
48. Phillips PE, Stuber GD, Heien ML, Wightman RE, Carelli RM. Subsecond dopamine release promotes cocaine seeking. Nature. 2003;422:614–8.

49. Kelley AE. Memory and addiction: shared neural circuitry and molecular mechanisms. Neuron. 2004;44:161–79.
50. Berke JD, Hyman SE. Addiction, dopamine, and the molecular mechanisms of memory. Neuron. 2000;25:515–32.
51. de Wit H. Impulsivity as a determinant and consequence of drug use: a review of underlying processes. Addict Biol. 2009;14:22–31.
52. Winstanley CA. The neural and neurochemical basis of delay discounting. In: Madden GJ, Bickel WK, editors. Impulsivity: the behavioral and neurological science of discounting. Washington DC: American Psychological Association; 2010.
53. Cardinal RN, Pennicott DR, Sugathapala CL, Robbins TW, Everitt BJ. Impulsive choice induced in rats by lesions of the nucleus accumbens core. Science. 2001;292:2499–501.
54. Winstanley CA, Bachtell RK, Theobald DE, Laali S, Green TA, Kumar A, Chakravarty S, Self DW, Nestler EJ. Increased impulsivity during withdrawal from cocaine self-administration: role for DeltaFosB in the orbitofrontal cortex. Cereb Cortex. 2009;19:435–444.
55. Pattij T, Vanderschuren LJ. The neuropharmacology of impulsive behavior. Trends Pharmacol Sci. 2008;29(4):192–99.
56. Volkow ND, Fowler JS, Wang GJ, Hitzemann R, Logan J, Schlyer DJ, Dewey SL, Wolf AP. Decreased dopamine D2 receptor availability is associated with reduced frontal metabolism in cocaine abusers. Synapse. 1993;14:169–77.
57. Franklin TR, Acton PD, Maldjian JA, Gray JD, Croft JR, Dackis CA, O'Brien CP, Childress AR. Decreased gray matter concentration in the insular, orbitofrontal, cingulate, and temporal cortices of cocaine patients. Biol Psychiatry. 2002;51(2):134–42.
58. Winstanley CA, Dalley JW, Theobald DEH, Robbins TW. Fractionating impulsivity: contrasting effects of central 5-HT depletion on different measures of impulsive behavior. Neuropsychopharmacology. 2004;29:1331–43.
59. Mobini S, Body S, Ho MY, Bradshaw CM, Szabadi E, Deakin JFW, Anderson IM. Effects of lesions of the orbitofrontal cortex on sensitivity to delayed and probabilistic reinforcement. Psychopharmacology. 2002;160:290–98.
60. Gipson CD, Lee DC, Kelly TH, Kalivas PW, Bardo MT. Impulsivity as a risk-related multifaceted construct relevant to drug abuse: a translational and neurobiological review. (in preparation).
61. Cardinal RN, Parkinson JA, Lachenal G, Halkerston KM, Rudarakanchana N, Hall J, Morrison CD, Howes SR, Robbins TW, Everitt BJ. Effects of selective excitotoxic lesions of the nucleus accumbens core, anterior cingulate cortex, and the central nucleus of the amygdala on autoshaping performance in rats. Behav Neurosci. 2002;116:533–67.
62. Hall J, Parkinson JA, Connor TM, Dickinson A, Everitt BJ. Involvement of the central nucleus of the amygdala and nucleus accumbens core in mediating Pavlovian influences on instrumental behavior. Eur J Neurosci. 2001;13:1984–92.
63. Acheson A, Farrar AM, Patak M, Hausknecht KA, Kieres AK., Choi S, de Wit H, Richards JB. Nucleus accumbens lesions decrease sensitivity to rapid changes in the delay to reinforcement. Behav Brain Res. 2006;173:217–28.
64. Winstanley CA, Theobald DEH, Cardinal RN, Robbins TW. Contrasting roles for basolateral amygdala and orbitofrontal cortex in impulsive choice. J Neurosci. 2004;24:4718–22.
65. Koob GF, Ahmed SH, Boutrel B, Chen SA, Kenny PJ, Markou A, O'Dell LE, Parsons LH, Sanna PP. Neurobiological mechanisms in the transition from drug use to drug dependence. Neurosci Biobehav Rev. 2004;27:739–49.
66. Solomon RI, Corbit JD. An opponent-process theory of motivation: I. Temporal dynamics of affect. Psychol Rev. 1974;81(2):119–45.
67. O'Brien CP. A range of research-based pharmacotherapies for addiction. Science. 1997;278(5335):66–70.
68. Winstanley CA, Olaussen P, Taylor JR, Jentsch JD. Insight into the relationship between impulsivity and substance abuse from studies using animal models. Alcohol: Clin Exp Res. 2010;34:1–13.

69. Grimm JW, Hope BT, Wise RA, Shaham Y. Neuroadaptation. Incubation of cocaine craving after withdrawal. Nature. 2001;412:141–2.
70. Shaham Y, Shalev U, Lu L, de Wit H, Stewart J. The reinstatement model of drug relapse: history, methodology, and major findings. Psychopharmacology (Berl). 2003;168(1–2):3–20.
71. McFarland K, Davidge SB, Lapish CC, Kalivas PW. Limbic and motor circuitry underlying footshock-induced reinstatement of cocaine-seeking behavior. J Neurosci. 2004;24(7): 1551–60.
72. Grimm J, See R. Dissociation of primary and secondary reward relevant limbic nuclei in an animal model of relapse. Neuropsychopharmacology. 2000;22:473–9.
73. Meil WM, See RE. Lesions of the basolateral amygdala abolish the ability of drug associated cues to reinstate responding during withdrawal from self-administered cocaine. Behav Brain Res. 1997;87(2):139–48.
74. Bava S, Tapert SF. Adolescent brain development and the risk for alcohol and other drug problems. Neuropsychol Rev. 2010;20:398–413.
75. Geidd JN. The teen brain: insights from neuroimaging. J Adolescent Health. 2008;42(4): 335–43.
76. Spear LP. The adolescent brain and age-related behavioral manifestations. Neurosci Biobehav Rev. 2000;24:417–63.
77. Schell TL, Orlando M, Morral AR. Dynamic effects among patients' treatment needs, beliefs, and utilization: a prospective study of adolescents in drug treatment. Health Serv Res. 2005;40(4):1128–47.
78. Cyders MA, Smith GT. Emotion-based dispositions to rash action: positive and negative urgency. Psychol Bull. 2008;134(6):807–28.
79. Dickman SK. Functional and dysfunctional impulsivity: personality and cognitive correlates. J Pers Soc Psychol. 1990;58:95–102.
80. Gould E, Woolf NJ, Butcher LL. Postnatal development of cholinergic neurons in the rat: I. Forebrain. Brain Res Bull. 1991; 27:767–789.
81. Kostovic I. Structural and histochemical reorganization of the human prefrontal cortex during perinatal and postnatal life. Prog Brain Res. 1990;85:223–39.
82. Woo TU, Pucak ML, Kye CH, Matus CV, Lewis DA. Peripubital refinement of the intrinsic and associational circuitry in monkey prefrontal cortex. Neuroscience. 1998;80(4):1149–58.
83. Brenhouse HC, Sonntag KC, Andersen SL. Transient D1 dopamine receptor expression on prefrontal cortex projection neurons: relationship to enhanced motivational salience of drug cues in adolescence. J Neurosci. 2008;28:2375–82.
84. Brenhouse HC, Dumais K, Andersen SL. Enhancing the salience of dullness: behavioral and pharmacological strategies to facilitate extinction of drug-cue associations in adolescent rats. Neuroscience. 2010;169(2):628–36.
85. Goldstein RA, Volkow ND. Drug addiction and its underlying neurobiological basis: neuroimaging evidence for the involvement of the frontal cortex. Am J Psychiatry. 2002;159: 1642–52.
86. Cressman VL, Balaban J, Steinfeld S, Shemyakin A, Graham P, Parisot N, Moore H. Prefrontal cortical inputs to the basal amygdala undergo pruning during late adolescence in the rat. J Comp Neurol. 2010;518(14):2693–709.
87. Doremus-Fitzwater TL, Varlinksaya EI, Spear LP. Motivational systems in adolescence: possible implications for age differences in substance abuse and other risk-taking behaviors. Brain Cogn. 2010;72:114–23.
88. Li C, Frantz KJ. Attenuated incubation of cocaine seeking in male rats trained to self-administer cocaine during periadolescence. Psychopharmacology (Berl). 2009;204(4):725–33.
89. Anker JJ, Carroll ME. Reinstatement of cocaine seeking induced by drugs, cues, and stress in adolescent and adult rats. Psychopharmacology (Berl). 2010;208(2):211–22.
90. Arnett J. Reckless behavior in adolescence: a developmental perspective. Dev Rev. 1992;12:339–73.

91. Gardner M, Steinberg L. Peer influence on risk taking, risk preference, and risky decision making in adolescence and adulthood: an experimental study. Dev Psychol. 2005;41(4): 623–35.
92. Gipson CD, Yates J, Beckmann JS, Marusich JA, Bardo MT. Social facilitation of d-amphetamine self-administration in rats. (in review).
93. Douglas LA, Varlinskaya EI, Spear LP. Rewarding properties of social interactions in adolescent and adult male and female rats: impact of social versus isolate housing of subjects and partners. Dev Psychobiol. 2004;45:153–162.
94. Varlinskaya E, Spear LP. Changes in sensitivity to ethanol-induced social facilitation and social inhibition from early to late adolescence. Ann N Y Acad Sci. 2006;1021(1):459–61.

Cassandra D. Gipson Postdoctoral Fellow in Neuroscience and Researcher at the Department of Neuroscience, University of South Carolina, School of Medicine, USA.

Peter W. Kalivas Neuroscientist. Honorary Professor, University of Córdoba, Argentina. Honorary Professor, Department of Neuroscience, School of Medicine, Nanjing University, China. Professor, Department of Neuroscience, University of South Carolina, School of Medicine, USA. Member, Research Council, National Alliance for Research in Schizophrenia and Affective Disorders. PhD, University of Seattle, USA. Postdoctoral fellowship at the University of North Carolina, USA.

Chapter 5
Neurobiology of the Action of Drugs of Abuse

Roseli Boerngen-Lacerda

Adolescence, the transition from childhood to adulthood, is characterized by remarkable biological, cognitive, emotional and social changes. In addition, this is the stage of life when the use of drugs usually begins. That fact makes patent the relevance of studying the processes and mechanisms involved in the development of the central nervous system (CNS) that predispose individuals to drug use as well as of understanding the potential alterations associated with drug use in adolescence that might increase the users' vulnerability to future dependence. Although the developing brain is quite resilient to neurotoxicity, exposure to drugs in critical periods of neurological development might interrupt the normal course of the brain's maturation and thus interfere with processes essential for adult cognitive functions, including negative effects on school and work performance as well as on various social activities, i.e., persistent alterations of the user's life, even after having quit the drug. The use of drugs in adolescence is associated with abnormal changes at various levels of the brain's structure, including its molecular, genetic and epigenetic aspects, neural circuits, cellular structure and gray-white-matter ratio, as well as in brain functions, such as neurocognition, verbal skills and impulse control [1, 2].

Studies conducted with neuroimaging methods have shown that the CNS is still undergoing development in early adulthood, i.e., in the late twenties [3]. For instance, approximately 50 % of the neurons in the prefrontal cortex, which is associated with impulse control and other executive functions, are still subject to modeling at that age, which might account for the high vulnerability of adolescents to the effects of drugs. Adolescent rats had lower sensitivity to the unpleasant effects of ethanol (e.g., vomiting, "hangover") and higher sensitivity to ethanol-induced reduction of social inhibition compared to older rats. In addition, the younger animals also exhibit lower sensitivity to the anxiolytic effect of alcohol [4]. These findings together might indicate that adolescents have diminished ability to moderate their ethanol intake. Other studies also conducted with adolescent rats have shown that chronic exposure to ethanol induces greater cognitive impairment compared

R. Boerngen-Lacerda (✉)
Department of Pharmacology, Federal University of Paraná (UFPR), Curitiba, Brazil
e-mail: boerngen@ufpr.br; boerngen@hotmail.com

D. De Micheli et al. (eds.), *Drug Abuse in Adolescence*,
DOI 10.1007/978-3-319-17795-3_5

to adult rats subjected to a similar amount of alcohol [5]. In humans, the hippocampus is smaller and the memory retention for verbal and non-verbal tasks lower among adolescents exhibiting a "heavy drinking" pattern [6]. Although it is not possible to establish with full certainty whether such alterations are the cause or the consequence of alcohol consumption, there is, nonetheless, an evident relationship between the pattern of alcohol consumption and cognitive disorders in adolescents.

In adolescence, the myelination of the frontal lobe is associated with specific behaviors that are normal at that stage, such as exposure to risk, the search for novelties and giving in to peer pressure, which make adolescents increasingly likely to try legal and illegal drugs [7]. When first trying drugs, adolescents experience transient pleasurable effects, changes in their mental states and apparent improvement of performance, which might lead them to use drugs more often. Those effects are known as the reinforcing properties of drugs and serve as the initial motivations for using them. To feel those effects again, people start using drugs repeatedly and may eventually become dependent. Dependence is defined as an "intense desire for the drug with an impaired ability to control the urges to take that drug, even at the expense of serious adverse consequences" [8]. To summarize, individuals first experience the reinforcing properties of drugs while keeping control of their powers of decision-making; however, when they become dependent, they lose control over their own choices. That transition from recreational use to drug dependence is associated with changes in gene expression and protein production in cells, resulting in alterations of neuronal circuits in structures associated with motivational behavior, executive functions and inhibitory control [9]. Those are precisely the brain areas that undergo development during adolescence.

In adolescence, the brain structures change through a process of synaptic refinement and myelination. Synaptic refinement involves the removal of unnecessary connections (pruning) and the strengthening of the well-used ones (synaptogenesis). Those two mechanisms are controlled by the environmental experiences, resulting in reduction of the gray matter, especially in the prefrontal and temporal cortices and subcortical structures, such as the thalamus, striatum and nucleus accumbens. Myelination, in turn, increases the integration, speed and efficiency of the neural conductivity. In the normal course of brain development, the higher association areas mature only after the lower sensorimotor ones have completed their development, with frontal lobes being the last to achieve maturity. Concomitant myelination improves the efficiency of the communication between the frontal and the subcortical areas, allowing for better control of the executive and inhibitory cognitive functions [1]. Consequently, the influences to which youths are exposed are determinant for the process of maturation.

The alteration in the CNS that takes place in adolescence also occurs in the neurotransmitter systems, many of which are direct or indirect targets of drugs [10]. For instance, excitatory synapses, such as the glutamatergic ones, are eliminated from the prefrontal cortex in adolescence, whereas the dopaminergic projections to the prefrontal cortex increase [2] In adolescent rats, the D1, D2 and D4 receptors increase in the frontal and entorhinal cortices and hippocampus, and the D1 and D2 receptors and dopamine turnover increase in subcortical structures, such as the striatum and nucleus accumbens [11]. At the cellular level, continuous brain reorganization takes place in adolescence and early adulthood as a result of the interaction between the cAMP response element-binding (CREB) protein transcription factor, which plays a relevant role in postnatal neurochemical remodeling, and the growth factor BDNF (*brain derived neurotrophic factor*), which participates in neuronal differentiation and maintenance and in neuroplasticity [10]. The changes that occur during adolescence may interact with the actions of drugs. For instance, in sections of the hippocampus of rats, the inhibitory action of alcohol on the synaptic potentials mediated by the N-methyl-D-aspartate (NMDA) glutamatergic receptor and long-term potentiation (LTP) have been observed to be higher in adolescents compared with adults [12]. Chronic exposure to alcohol has been shown to increase the size of the dendritic spines, which might account for the neuroplastic changes related to the alcohol-related learning that stabilizes addictive behaviors in adulthood [13]. The use of histological and electrophysiological techniques has shown that the brains of adolescent rats are more sensitive to the following effects of alcohol: intoxication-induced neurodegeneration [14], inhibition of neurogenesis [10], increase of GABAergic neurotransmission [15], spontaneous interneuron firing [16] and increases in serotonergic transporter density in several brain areas [17]. As mentioned above, adolescent animals are less sensitive to the depressor effects of alcohol, such as sedation and motor impairment, and to the effects of acute abstinence (hangover) [2]. Those differences may thus be related to the glutamatergic and GABAergic systems. Electrophysiological studies with laboratory animals have found differences between adolescent and adult rats with regard to the alcohol-induced inhibitory responses mediated by the GABA-A receptor [18]. Adolescent rats are not sensitive to the effects of acute ethanol withdrawal syndrome, such as anxiety, seizures, and suppression of activity. However, after repeated withdrawal episodes, anxiety behavior increases and, the seizure threshold decreases for both adults and adolescents [19].

Two mechanisms have been proposed to account for the neurotoxicity of alcohol [5]. According to one such mechanism, intermittent use of alcohol may induce short withdrawal episodes that, in turn, cause neuronal damage by means of glutamatergic receptor (NMDA)-mediated excitotoxicity. Chronic adaptations of those receptors would thus account for actual structural damage and the onset of a more complicated withdrawal syndrome that includes seizures. Such adaptations have actually been observed, mainly in studies conducted with animals and isolated neurons. However, De Bellis et al. [20] reported that out of 14 adolescents with alcohol-related disorders and reduction of the prefrontal cortex, only two exhibited alcohol withdrawal symptoms. The second mechanism invoked to account for the neurotoxicity of alcohol is related to the ability of that substance to induce neuroinflammation. Alcohol consumption induces the release of inflammatory mediators in the brain through activation of the glial cells and stimulation of intracellular signaling pathways that promote induction of the proinflammatory cytokines interleukin (IL)-1ß and tumor necrosis factor (TNF)-α, cyclooxygenase-2 (COX-2), inducible nitric oxide synthase (iNOS) and cell death. In animals given alcohol during adolescence, increases of the inflammatory mediators in the brain occur concomitant with short- and long-term cognitive impairment. Administration of a single large dose of ethanol in mid-adolescence was shown to reduce neuronal proliferation and survival in young adult animals. Those findings suggest that the drinking pattern characterized by episodes of intoxication induces long-lasting effects on neurogenesis with regard to learning and other behaviors that undergo maturation during adolescence. The two mechanisms just described have been suggested to account for the neurotoxicity of alcohol, but few studies have sought to assess the neurotoxicity of other drugs [5]. Hernandez-Rabaza et al. [21] subjected adolescent rats to repeated treatment with cocaine or 3,4-methylenedioxy-methamphetamine (MDMA or ecstasy) alone or combined with alcohol and assessed the effects of those substances on memory task performance, hippocampal neurogenesis in adults and the degree of neurotoxicity. The results showed that persistent memory impairment and neurotoxicity in the dentate gyrus, characterized by a reduction in the number of granule neurons, only occurred when ecstasy was used in combination with alcohol. A study that compared young individuals with marijuana dependence and healthy subjects found that impaired learning and performance in attention tasks were related to a reduction in plasma NGF (*nerve growth factor*) levels but did not find any change in the BDNF levels [22]. The authors of that study suggested that a reduction in NGF levels may explain the neurotoxicity of marijuana, representing a risk factor for the onset of psychosis.

One further negative consequence of exposure to alcohol in adolescence is an increased risk for abuse and dependence in adulthood. That fact is supported by several pieces of evidence emerging from studies with animals [2]; however, few studies have yet been conducted with humans in this regard [23]. Studies conducted with C57BL/6J and BALB mice offered a free choice of water or alcohol immediately after weaning found increased consumption of and preference for alcohol in adulthood. In rats exposed to alcohol in adolescence, the greatest consumption of and preference for alcohol in adulthood was found to be associated with reduced dopamine receptor D2 (DRD2) protein levels and with phosphorylation of the glutamatergic receptor subunit 2B (NMDAR2B) in brain areas relevant to dependence (e.g., the prefrontal cortex, the hippocampus). Chronic use of alcohol was found to induce a steady release of dopamine in the nucleus accumbens, and its levels were much higher in the adolescent compared with the adult rats. In addition, changes were found in chromatin remodeling due to altered acetylation of histones H3 and H4 in several brain areas (prefrontal cortex, nucleus accumbens, striatum), which is an epigenetic mechanism known to play a highly relevant role in learning processes and that might be directly related to increased vulnerability to alcohol dependence in adulthood. Further studies are needed to establish which genes are modulated by means of acetylation or methylation and which genes participate in the long-lasting neurobehavioral disorders induced by alcohol consumption in adolescence [5].

Drugs of abuse are substances or a set of substances that do not bear any similarity in regard to their chemical structure or their primary acute mechanisms of action. According to their main acute effects on the CNS, they might be classified as follows:

- depressants (e.g., ethanol, inhalants, sedatives, benzodiazepines, opioids);
- stimulants (e.g., amphetamines, cocaine, nicotine, xanthine derivatives); and
- hallucinogens (e.g., lysergic acid diethylamide [LSD], cannabinoids [marijuana], methamphetamine derivatives [methyilenedioxy-methamphetamine or *ecstasy*], ketamine, psilocybin, anticholinergic agents).

The molecular targets for the primary actions of those drugs in the CNS are described in Table 5.1, and the resulting acute and chronic effects are described in Table 5.2. Due to their primary actions, these drugs act on various brain structures that account for their addictive properties. All the drugs of abuse induce acute increases in the dopamine concentration in the nucleus accumbens (also known as the ventral striatum), which receives projections from the mesencephalon (ventral tegmental area—VTA). Moreover, other brain areas, such as the amygdala, the prefrontal cortex and other cortical limbic areas (e.g., the hippocampus and hypothalamus) interact with that circuit. All those structures are components of the brain's reward system, which is activated by natural reinforcers, such as food, sex and social interaction [9]. Increased dopamine release is not directly related to reward as such but instead with the prediction [24] and motivational salience [25] of reward. Motivational salience alludes to the changes in stimuli or the environment that trigger alertness or attention behaviors. Thus, salience might be related not only to reward but also to aversive or unexpected stimuli and to novelty, thus playing a relevant role in conditioned learning. In conditioned learning, the drug-induced dopamine increase is associated with stimuli that were originally neutral. As a consequence of that association, those stimuli become able to increase the dopamine release and to elicit craving for the drug, i.e., they acquire motivational salience [25]. This phenomenon explains why addicted individuals are at high risk of relapse when they are exposed to environments or stimuli previously associated with the use of the drug. Also relevant are the qualitative and quantitative differences in the dopamine release induced by natural reinforcers or drugs. The amount of drug-induced dopamine release is larger (approximately five to ten times) and its duration longer compared with natural reinforcers [26]. In addition, habituation occurs in the case of the increased dopamine release induced by natural reinforcers but not by drugs [27], which suggests that the motivational properties of drugs are intensified by chronic use.

Dependence-related adaptations occur at various levels. At the cellular level, chronic use of drugs modifies the expression of transcription factors (e.g., Δ-FosB, CREB) in the brain areas regulated by dopamine [25]. Those factors modulate the synthesis of proteins involved in the synaptic plasticity that accounts for the morphological changes in the neurons from dopamine-regulated circuits (such as proliferation of the dendritic tree and density of synaptic spines) [9]. Several other neurotransmitters are also affected, such as glutamate, GABA, opioids, serotonin and various neuropeptides. In individuals addicted to cocaine, neuroimaging methods found that the number of dopaminergic D2 receptors is reduced in the orbitofrontal cortex and anterior cingulate gyrus, which are areas related to compulsive drug-taking behavior [8]. Moreover, brain circuits undergo adaptations in response to the chronic use of drugs. Changes in the mesocortical circuit, including the

Table 5.1 Main mechanisms of action of the main psychotropic drugs. (Adapted from Katzung BG. Basic and clinical pharmacology, 10. ed. Norwalk: Appleton and Lange; 2007)

Drug	Main molecular target	Action	Effect on dopaminergic neurons
Drugs that bind to ionotropic receptors and ion channels			
Nicotine	nAChR	Agonist	Disinhibition
Alcohol	$GABA_AR$; 5-HT_3R; nAChR; NMDAR; Kir3 channels	–	Disinhibition
Benzodiazepines	$GABA_AR$	Positive modulator	Disinhibition
Phencyclidine, ketamine	NMDAR	Antagonist	–
Drugs that activate protein G-coupled receptors			
Opioids	OR (G_{io})	Agonist	Excitation, disinhibition
Cannabinoids	CB1R (G_{io})	Agonist	Excitation, disinhibition
Gamma-hydroxybutyric acid (GHB)	$GABA_BR$ (G_{io})	Weak agonist	Disinhibition
LSD, mescaline, psilocybin	5-$HT_{2A}R$ (G_q)	Partial agonist	–
Drugs that bind to biogenic amine transporters			
Cocaine	DAT, SERT, NET	Inhibitor	Reuptake blockade
Amphetamines	DAT, NET, SERT, VMAT	Reverse transport	Reuptake blockade, depletion
Methyilenedioxy-methamphetamine (*ecstasy*)	SERT>DAT, NET	Reverse transport	Reuptake blockade, depletion

5-HT_XR various serotonin receptor subtypes, *CB1R* endocannabinoid receptor type 1, *DAT* dopamine transporter, *$GABA_XR$* various gamma-aminobutyric acid receptor subtypes, *Kir3* G protein-coupled potassium channel, inwardly rectifying, *LSD* lysergic acid diethylamide, *OR* opioid receptor, *nAChR* nicotinic acetylcholine receptor, *NET* norepinephrine transporter, *NMDAR* glutamate N-methyl-D-aspartate receptor, *SERT* serotonin transporter, *VMAT* vesicular monoamine transporter

Table 5.2 Main acute and chronic effects of the main psychotropic drugs

Drug	Main acute effects on the CNS	Main chronic effects (CNS and periphery)
Nicotine	Euphoria, anxiolysis	Cardiovascular diseases and stroke; chronic obstructive pulmonary disease; kidney disease; lung, bladder, breast, oral, throat and esophageal cancer; worsening of diabetes, asthma and hypertension; tolerance and dependence
Alcohol	Euphoria, aggressiveness, sedation, motor incoordination, ataxia, coma and death (according to dose)	Hypertension and stroke; anxiety, depression and suicide; liver disease; gastric ulcer and pancreas inflammation; episodes of amnesia and hallucinations; memory and cognitive impairment; impotence; premature aging; permanent brain lesions resulting in memory loss, disorientation and cognitive deficit; difficult mobility due to osteoporosis, gout and neuromuscular affection; oral, throat and breast cancer; tolerance and dependence
Benzodiazepines and other hypnotic sedatives	Sedation, anxiolysis, muscle relaxation	Sleepiness, dizziness and confusion; gait disorders and falls; depression; sleep disorders; headache; skin rashes; nausea; tolerance and dependence
Phencyclidine, ketamine	Euphoria/sedation, hallucinations	Attention, learning and memory impairment; flashbacks; anxiety and depression.
Opioids (e.g., morphine, heroin)	Euphoria, analgesia, sedation, coma and death (according to dose)	Loss of teeth; severe constipation; irregular menstrual periods; impotence and loss of libido; tolerance and dependence
Cannabinoids (e.g., marijuana, hashish)	Euphoria/sedation, dysphoria, panic, hallucinations	Respiratory diseases; lung, respiratory and gastrointestinal cancer; hypertension; heart disease; asthma; bronchitis, emphysema; loss of memory and problem solving skills; loss of motivation; loss of libido; depression; tolerance and dependence
Gamma-aminobutyric acid (GHB)	Sedation, ataxia	Tolerance and dependence
Lysergic acid diethylamide (LSD), mescaline, psilocybin	Hallucinations (unpredictable content: varies among users and settings)	Difficulty sleeping; mood swings, anxiety, panic, paranoia or agitation; torpor, muscle weakness, tremor or psychomotor agitation; tachycardia and hypertension; aggravation of mental disorders
Cocaine	Euphoria, psychomotor agitation, anorexia	Difficulty sleeping; tachycardia; headache; weight loss; tingling; constant skin wounds; risk of accidents and injury; exhaustion and reduced immunity against infections; intense craving; paranoia; anxiety; depression; agitation and mania; aggressiveness; psychosis; sudden death by cardiovascular problems

Table 5.2 (continued)

Drug	Main acute effects on the CNS	Main chronic effects (CNS and periphery)
Amphetamines	Euphoria, psychomotor agitation, anorexia	Difficulty sleeping; loss of appetite and weight; dehydration; temporomandibular joint dysfunction; headache and muscle pain; tremor and irregular heartbeats; reduced immunity against infections; psychosis; mood swings, including anxiety, depression, agitation and mania; hallucinations; aggressiveness
Methylenedioxy-methamphetamine (*ecstasy*)	Euphoria, hallucinations, hyperthermia	Hyperthermia; hydroelectrolytic imbalance; liver disorders; brain hemorrhage; amnesia; depression; panic; flashbacks; hallucinations; irreversible brain damage

CNS central nervous system

orbitofrontal cortex and cingulate gyrus, are related to compulsive consumption and loss of inhibitory control, which are aspects relevant to relapse. The changes in the mesolimbic circuit, including the nucleus accumbens, amygdala and hippocampus, most likely cause increases in the motivational value (salience) of a drug and in the stimuli that are conditioned to it, parallel to the reduction of the sensitivity to natural reinforcers [25] The nigrostriatal circuit exhibits adaptations related to habituation and drug-use rituals [28].

All the aforementioned alterations involved in the development of dependence point to a process of learning and memory relative to the substance(s) used, and all the evidence currently available supports the cognitive-behavioral approach in the recovery of drug-dependent individuals. According to the literature, in the case of alcohol at least, "heavy use" in adolescence is associated with deleterious effects on neurocognitive function, affecting memory and the processing of information as concerns attention, as well as the speed of response, with consequent impairment of executive function [1]. Despite only a few reports having been published, some evidence indicates that long-lasting cognitive impairment follows the use of marijuana [29]. After 4 weeks of supervised abstinence, adolescents who smoked marijuana on a regular basis exhibited poorer performance in learning tests, cognitive flexibility, visual exploration and working memory.

Advances in neuroimaging techniques allow the detection of changes in the brain structures involved in drug-induced neurocognitive damage. However, those techniques are still seldom used; thus, the number of corresponding studies is small. Some evidence indicates that the severity of alcohol dependence or abuse is associated with a reduction of the left hippocampal volume, which might account for memory impairment [1]. Interestingly, the results of some studies [30] suggest that the concomitant use of alcohol and marijuana causes less impairment compared to the use of alcohol alone. The reason might be that those substances induce opposite adaptive responses (neuroinflammation and suppression of myelination), whereby the results of macromorphometric assessment appear normal. However, structural microanalysis shows that marijuana increases glial proliferation and white matter density while simultaneously reducing gray matter density, for which reason the hippocampal volume is normal in spite of the functional disorders [1]. Moreover, changes in the volume of the prefrontal cortex occur in adolescents exposed to "heavy use" of alcohol or marijuana, with females being particularly affected. Such changes are associated with the interruption of dendritic proliferation, which is associated with the impairment of verbal memory [31]. Maturation of the brain's white matter is relevant for the establishment of connections between brain areas. Few studies have addressed that subject, and the corresponding evidence does not indicate that either alcohol or marijuana induces changes in white matter volume. One single study has found that the use of marijuana and reduced white matter volume are equally predictive of the occurrence of depressive symptoms in adolescents [32]. Microstructural analysis of white matter has shown myelination abnormalities, especially in the corpus callosum, in adolescents with a "heavy use" pattern and episodes of alcohol intoxication, whereas concomitant use of marijuana has been shown to attenuate those alterations [33]. Cerebral blood flow is also a parameter deserving of attention because its reduction is associated with impaired oxygenation and tissue damage. Although stimulants cause cerebral vasoconstriction, no study investigating the possible differential effects of those drugs on adolescents can be found. One study [34] has found reduction of the blood flow in adolescent females with dependence ($n=8$) compared to "light drinkers" ($n=8$).

Given the characteristics inherent to adolescence, young people commonly try various substances, often in combination, to attain euphoria and pleasure, which increases the risk for abuse and dependence. The most typical combinations include alcohol and some other substance—marijuana, cocaine, an inhalant or a hallucinogen. The choice depends on the individual characteristics of each adolescent, the circumstances under which drug use occurs, and the price and availability of drugs [35].

One further risk factor in adolescence is the route of administration, which has influence on the potential for abuse and on the intensity of the effects of drugs. The time for onset and duration of the central effects of drugs depend on the route of administration as well as on the chemical and physical-chemical properties of drugs and their forms of presentation. In this regard, lipophilicity is crucial for drugs to affect the CNS. The potential for abuse increases when the onset of the central effects of a drug is quick (e.g., the euphoric effect of smoked crack cocaine starts in approximately 8 s, and that of intravenous cocaine starts in 14 s) and when their effect is transient. Various routes of administration are possible, including oral, intranasal, via other mucous membranes (vaginal, anal), respiratory (smoked, inhaled) and injection (subcutaneous, intramuscular, intravenous). With regard to the oral route, hypnotic sedatives and barbiturates are commonly used as tablets or pills, whereas cocaine and heroin might be used as powder or in liquid form. LSD is consumed by "licking" LSD-impregnated blotter paper. Marijuana and hashish might be added to food. When the oral route is used, absorption is slow, and the drug peak concentration is not high; consequently, the resulting euphoria is mild. The intranasal route is used with drugs such as cocaine, heroin and amphetamines. In that case, the solution must be water-soluble, absorption is low, and the plasma peak concentration is not high. The potential for abuse of the intranasal route is lower compared to the intravenous route and smoking. Drug intake through other mucous membranes, such as the vaginal and anal ones, is easy to hide; drugs such as cocaine and amphetamines might cause local injury due to their strong vasoconstrictor effect. Smoked drugs, including heroin/opium, crack/cocaine, marijuana/hashish, and even tobacco exhibit the highest potential for abuse. For instance, smoked crack and its related forms, "merla" and "oxi," induce intense euphoria in 8 s, concomitant with equally fast dysphoria, which stimulates additional drug-seeking through a feedback mechanism. Volatile substances, such as ether, chloroform, toluene, shoemaker's glue ("cola de sapateiro"), mixtures of chloroform/ether ("lança-perfume," "cheirinho da loló"), etc., are inhaled and do not need to be heated to be absorbed, in contrast to heroin and crack. Due to their wide availability and low cost, such substances are the most widely used by adolescents in Brazil, following alcohol and tobacco. Their potential for abuse is high because their pharmacokinetic profile is the same as that of smoked drugs. The intravenous route is not the most common among adolescents, but it exhibits high potential for abuse and additional risks, such as disease transmission through needles, including AIDS and hepatitis. That route is usually adopted at more advanced stages of drug dependence and demands more intensive and expensive therapeutic approaches, including hospitalization [35].

References

1. Squeglia LM, Jacobus J, Tapert SF. The influence of substance use on adolescent brain development. Clin EEG Neurosci. 2009;40(1):31–8.
2. Witt ED. Research on alcohol and adolescent brain development:opportunities and future directions. Alcohol. 2010;44:119–24.
3. Giedd J. Structural magnetic resonance imaging of the adolescent brain. In: Dahl RE, Spear LP, editors. Adolescent brain development: vulnerabilities and opportunities. New York: New York Academy of Sciences; 2004.
4. Varlinskaya EI, Spear LP. Acute effects of ethanol on social behavior and adult rats: role of familiarity of the test situation. Alcohol Clin Exp Res. 2002;26:1502–11.
5. Guerri C, Pascual M. Mechanisms involved in the neurotoxic, cognitive, and neurobehavioral effects of alcohol consumption during adolescence. Alcohol. 2010;44:15–26.
6. Tapert SF, Schweinsburg AD. The human adolescent brain and alcohol use disorders. In: Galanter M, editor. Recent developments in alcoholism. V. 17. Alcohol problems in adolescents and young adults: epidemiology, neurobiology, prevention, treatment. New York: Springer; 2005.
7. Sowell ER, Peterson BS, Thompson PM, et al. Mapping cortical change across the human life span. Nature Neurosci. 2003;6:309–15.
8. Volkow ND, Li T-K. Drug addiction: the neurobiology of behavior gone awry. In: Ries RK, Fiellin DA, Miller SC, et al., editors. Principles of addiction medicine. 4th edn. Philadelphia: Lippincott Williams and Wilkins; 2009.
9. Nestler EJ. From neurobiology to treatment: progress against addiction. In: Ries RK, Fiellin DA, Miller SC, et al., eds. Principles of addiction medicine. 4th edn. Philadelphia: Lippincott Williams and Wilkins; 2009.
10. Crews F, June H, Hodge C. Adolescent cortical development: a critical period of vulnerability to addiction. Pharmacol Biochem Behav. 2007;86:189–99.
11. Tarazi FI, Baldessarini RJ. Comparative postnatal development of dopamine D1, D2, and D4 receptors in rat forebrain. Int J Dev Neurosci. 2000;18:29–37.
12. White AM, Swartzwelder HS. Age-related effects of alcohol on memory and memory-related brain function in adolescents and adults. In: Galanter M, editor. Recent developments in alcoholism. V. 17. Alcohol problems in adolescents and young adults. New York: Kluwer Academic/Plenum Publishers; 2005.
13. Carpenter-Hyland EP, Chandler LJ. Adaptive plasticity of NMDA receptors and dendritic spines: implications for enhanced vulnerability of the adolescent brain to alcohol addiction. Pharmacol Biochem Behav. 2007;86:200–8.
14. Pascual M, Blanco AM, Cauli O, et al. Intermittent ethanol exposure induces inflammatory brain damage and causes long-term behavioural alterations in adolescent rats. Eur J Neurosci. 2007;25:541–50.
15. Fleming RL, Wilson W A, Swartzwelder HS. Magnitude and ethanol sensitivity of tonic GABAA receptor-mediated inhibition in dentate gyrus changes from adolescence to adulthood. J Neurophysiol. 2007;97:3806–11.
16. Yan H, Li Q, Fleming R, et al. Development sensitivity of hippocampal interneurons to ethanol: involvement of the hyperpolarization-activated current. J Neurophysiol. 2009;101:67–83.

17. Monti PM, Miranda R Jr, Nixon K, et al. Adolescence: booze, brains, and behavior. Alcohol Clin Exp Res. 2005;29:207–20.
18. Li Q, Wilson WA, Swartzwelder HS. Developmental differences in the sensitivity of spontaneous and miniature IPSCs to ethanol. Alcohol Clin Exp Res. 2006;30:119–26.
19. Wills TA, Knapp DJ, Overstreet DH, et al. Sensitization, duration, and pharmacological blockade of anxiety-like behavior following repeated ethanol withdrawal in adolescent and adult rats. Alcohol Clin Exp Res. 2009;33:455–63.
20. De Bellis MD, Narasimhan A, Thatcher DL, et al. Prefrontal cortex, thalamus, and cerebellar volumes in adolescents and young adults with adolescentonset alcohol use disorders and comorbid mental disorders. Alcohol Clin Exp Res. 2005;29:1590–600.
21. Hernandez-Rabaza V, Navarro-Mora G, Velazquez-Sanchez C, et al. Neurotoxicity and persistent cognitive deficits induced by combined MDMA and alcohol exposure in adolescent rats. Addict Biol. 2010;15(4):413–23.
22. Angelucci F, Ricci V, Spalletta G, et al. Reduced serum concentrations of nerve growth factor, but not brain-derived neurotrophic factor, in chronic cannabis abusers. Eur Neuropsychopharmacol. 2008;18(12):882–7.
23. Vetter CS, Doremus-Fitzwater TL, Spear LP. Time course of elevated ethanol intake in adolescent relative to adult rats under continuous, voluntary-access conditions. Alcohol Clin Exp Res. 2007;31:1159–68.
24. Schultz W, Tremblay L, Hollerman JR. Reward processing in primate orbitofrontal cortex and basal ganglia. Cereb Cortex. 2000;10:272–84.
25. Lu L, Grimm JW, Shaham Y, et al. Molecular neuroadaptation in the accumbens and ventral tegmental area during the first 90 days of forced abstinence from cocaine self-administration in rats. J Neurochem. 2003;85:1604–13.
26. Wise RA. Brain reward circuitry: insights from unsensed incentives. Neuron. 2002;36:229–40.
27. Di Chiara G. Nucleus accumbens shell and core dopamine: differential role in behavior and addiction. Behav Brain Res. 2002;137:75–114.
28. Porrino LJ, Lyons D, Smith HR, et al. Cocaine self-administration produces a progressive involvement of limbic, association and sensorimotor striatal domains. J Neurosci. 2004;24:3554–62.
29. Medina KL, Hanson K, Schweinsburg AD, et al. Neuropsychological functioning in adolescent marijuana users: Subtle deficits detectable after 30 days of abstinence. J Int Neuropsychol Soc. 2007a;13(5):207–20.
30. Medina KL, Schweinsburg AD, Cohen-Zion M, et al. Effects of alcohol and combined marijuana and alcohol use during adolescence on hippocampal volume and asymmetry. Neurotoxicol Teratol. 2007b;29:141–52.
31. Medina KL, McQueeny T, Nagel BJ, et al. Prefrontal cortex volumes in adolescents with alcohol use disorders: unique gender effects. Alcohol Clin Exp Res. 2008;32(3):386–94.
32. Medina KL, Nagel BJ, Park A, et al. Depressive symptoms in adolescents: associations with white matter volume and marijuana use. J Child Psychol Psychiatr. 2007c;48(6):592–600.
33. Jacobus J, McQueeny T, Bava S, et al. White matter integrity and neurocognition in adolescents with histories of binge drinking and marijuana use [paper]. Washington, DC: The Research Society on Alcoholism Scientific Conference; 2008.
34. Clark CP, Brown GG, Eyler LT, et al. Decreased perfusion in young alcohol-dependent women as compared with age-matched controls. Am J Drug Alcohol Abuse. 2007;33(1):13–9.
35. Estroff TW. Routes of abuse and specific drugs. In: Estroff TW, editor. Manual of adolescent substance abuse treatment. Washington, DC: American Psychiatric Publishing; 2001.

Roseli Boerngen-Lacerda Professor, Department of Pharmacology, Federal University of Paraná (UFPR). Lecturer, Master and Doctoral Program in Pharmacology, UFPR. DSc, Department of Psychobiology, Federal University of São Paulo (UNIFESP). BS in Biomedicine and MS in Pharmacology, UNIFESP.

Chapter 6
Neuroimaging of the Human Brain in Adolescent Substance Users

Michael Takagi, George Youssef and Valentina Lorenzetti

Introduction

Substance misuse represents one of the most problematic causes of adolescent morbidity and mortality worldwide [1, 2]. The initiation of substance misuse typically occurs during the adolescent period, with data revealing that as many as 48 % of students aged 18 years report lifetime substance use [3]. Amongst US adolescents, the most common substances misused include alcohol, cannabis, and inhalants (e.g., glue, spray paint, gasoline) [3–5]. These statistics are alarming and underscore an urgent need to characterize the adverse outcomes associated with adolescent substance misuse in order to inform policies aimed at reducing the burden of these behaviors.

Whilst scientific investigation into drugs of abuse is centuries old [6] the recent emergence of neuroimaging methodologies has heralded a new era of investigation by illuminating how substance misuse can impact neurobiology and cognitive functioning. Importantly, a growing body of imaging research has increasingly reported cognitive and neurobiological harms associated with the initiation and the

M. Takagi (✉)
Clinical Sciences, Murdoch Childrens Research Institute, Melbourne, Australia

School of Psychological Sciences, University of Melbourne, Melbourne, Australia
e-mail: michael.takagi@mcri.edu.au

G. Youssef
School of Psychology, Deakin University, Australia

Monash Clinical and Imaging Neuroscience, School of Psychological Sciences and Monash Biomedical Imaging, Monash University, Australia

V. Lorenzetti
Monash Clinical and Imaging Neuroscience, School of Psychological Sciences & Monash Biomedical Imaging, Monash University, Clayton, VIC Australia

Melbourne Neuropsychiatry Centre, Department of Psychiatry, The University of Melbourne, Parkville, VIC Australia

D. De Micheli et al. (eds.), *Drug Abuse in Adolescence*,
DOI 10.1007/978-3-319-17795-3_6

consumption of drugs during adolescence [1, 7, 8]. Also, the adverse outcomes associated with substance misuse in adolescence have been shown to be more severe than those reported in to adulthood [1]. This has been particularly noted for those who initiate substance use before 17 years of age [1, 7, 8]. This susceptibility to negative outcomes is hypothesized to be a consequence of interactions between drug effects and the developing adolescent brain.

From this basis, this chapter will provide insights into how neuroimaging research has informed research into the adverse neurobiological and cognitive outcomes of adolescent substance misuse. Specifically, we will focus on studies that employ Magnetic Resonance Imaging (MRI) techniques, which use powerful magnetic field and radio waves to provide high-resolution images of brain structure and function (e.g., grey/white matter volume, microstructure integrity of brain tissue, and activation of neural networks) [9]. Since its first use in the early 1970s, numerous imaging techniques and analysis methods have been developed to examine brain tissue in fine detail. However, only more recently have significant advances in neuroimaging technology been applied to addiction models and examinations of the consequences of adolescent substance misuse been conducted in earnest.

This chapter presents a cross-section of studies that have utilized MRI techniques to examine the impact of substance use on the adolescent brain. Specifically, we focus on examining the neurobiological outcomes associated with regular exposure to cannabis, alcohol, and inhalants, the three most commonly used (and investigated) substances of use in adolescence [4, 5].

The Adolescent Period

Adolescence can be formally defined as the developmental period involving the transition from childhood, to adult-like biological, psychological and social maturity [10, 11]. Based on this formal definition, adolescence ends when the individual becomes an "adult", although this is difficult to define since adolescent maturational processes continue into the third decade of life [12–14]. In this review, we refer to adolescence as the period occurring approximately between 13 and 18 years of age as evidence suggests that drug exposure during this period of adolescence (relative to later adolescence/early adulthood) is associated with more serious neurobiological consequences for the individual [1, 7, 8].

Adolescent Brain Development, Cognition and Substance Use

Relevant for the study of drugs of abuse is the significant brain remodeling that occurs throughout the adolescent period, characterized by increased development of white matter tracts (i.e., myelination) and decreases in cortical gray matter through synaptic pruning [13]. The increased myelination and elimination of excess synaptic connections through apoptotic processes results in refinement of the neural circuitry. This, in turn, strengthens the remaining functional connections and reduces competition from suboptimal associations. As a result, communication across distributed systems and associated cognitive, emotional and social processing is vastly improved.

Within the context of general myelination and synaptic pruning processes are regional differences in the rate of brain maturation during this period that may have particular relevance for the understanding of adolescent substance use. For example researchers have highlighted the observed differences in maturational trajectory of brain systems involved in reward processing (typically matures earlier in adolescence and associated with sensation and novelty seeking) and the maturation of systems involved in cognitive, emotional, and behavioural regulation (typically later in adolescence) as being a potential cause for adolescent vulnerability to substance misuse [15, 16]. These regulatory processes, known as executive functioning, comprises a number of abilities (such as sustained attention, impulse control, working memory, cognitive flexibility, planning, decision making and cognitive control) that work in an integrated manner to allow an individual to engage in purposeful, goal-directed behaviour [17]. Consequently, adolescent developmental processes may predispose these individuals to engage in substance use behaviors through an increased motivation towards the rewarding aspects of drugs of abuse coupled with an immature ability to regulate these maladaptive behaviours.

Key Brain Systems Relevant to Adolescent Substance Misuse

Given the putative importance that differences in the rate of maturation of reward and executive functioning circuits may have in predisposing adolescents in particular to substance misuse, much of the neuroimaging research has focused on investigating the structure and function of these brain systems. Specifically, reward and affective neurocircuitry include the ventral striatum (of which the nucleus accumbens is the most prominent region), dorsal striatum, amygdala, hippocampus, ventral anterior cingulate cortex (vACC), ventromedial prefrontal cortex (vmPFC) and orbitofrontal cortex (OFC) which are all heavily innervated with dopamine neurons [18–20]. In contrast, two brain regions consistently implicated in executive functioning and that develop later in adolescence include the dorsolateral PFC and dorsal ACC [17]. Additionally, researchers have targeted brain regions ascribed to the regulation of cognitive and emotional behaviors, the alteration of which has been reported in substance using populations. These include regions such as: (i) the hippocampus, which plays a key role in memory and learning processes; (ii) components of the hypothalamic-pituitary-adrenal axis (e.g. pituitary gland and hormonal stress levels), the main biological substrate of the stress response; (iii) the cerebellum, which is involved in mediating higher-order executive functions and motor control; and (iv) white matter fiber tracts, that enables structural connectivity between key brain regions (e.g., fronto-temporal, fronto-cerebellar).

There are numerous animal studies that have investigated the effects of drugs on these brain regions, with evidence suggesting that substance use may significantly impair adolescent cognitive and brain development [21–23]. For example, in reviewing the animal evidence, Smith [24] reported that exposure to substances in adolescent rats may disrupt normal pubertal development, potentially leading to abnormal CNS development during adolescence and changes in adult cognitive capabilities. Further, Smith (2003) reported that adolescent substance misuse has a larger effect on cognition when compared adult-onset substance misuse. These findings mirror the preliminary findings in humans [20].

Broadly, the animal evidence supports the notion that adolescent substance misuse is associated with marked neurobiological and cognitive abnormalities that can be observed in adulthood. However, only a limited number of investigations have been conducted in adolescent samples specifically (i.e., samples with a mean age of 18 years). This may be due to the relatively recent advances and increased availability of imaging techniques. Additionally, obtaining ethics approval for adolescent participants typically requires an added level of parental consent which may explain the relatively small number of investigations with adolescent samples. As a result, our understanding of the adverse neurobiological outcomes of adolescent substance use is limited.

To summarise, the adolescent brain is undergoing substantial development during adolescence. This has important implications for understanding how substance use affects neurobiological and cognitive development. Specifically, evidence has demonstrated that the neurocircuitry underlying reward processing develops early in adolescence whilst the development of prefrontal areas important for executive function and behavioural regulation has a protracted developmental trajectory. This disjointed development may predispose adolescents to seek the rewarding aspects drugs of abuse whilst having an immature ability to regulate behaviour appropriately. Thus, adolescents may be particularly susceptible to possible adverse effects of substance use during this period. Consequently, neuroimaging research has focused predominately on key brain regions that are ascribed to the regulation of reward, emotional processing and executive functions; which are necessary to achieve adult level behavioural and emotional regulation. From this context, the remainder of this chapter will illustrate how neuroimaging methods have provided insights into how drugs can affect the adolescent brain and cognition.

Method

To provide a context for interpreting imaging findings on adolescent substance use, we characterised the studies emerging from the MRI literature to date. Studies were reviewed by employed imaging technique (i.e., structural Magnetic Resonance Imaging (sMRI), Diffusion Tensor Imaging (DTI), functional MRI (fMRI)). Samples were then characterised with respect to demographic, matching criteria and levels of exposure, while the imaging findings were reviewed by substance of exposure (i.e., cannabis, alcohol, inhalant, and other substances).

Search Method

An extensive search of PubMed was conducted to identify imaging studies investigating adolescent substance users (SUD). We utilised the following search terms: (inhalant OR "cannabis" OR "marijuana" OR "alcohol" OR ecstasy OR amphetamine OR cocaine OR Heroin OR opiate) AND (MRI OR DTI OR fMRI) AND (Humans[Mesh] AND adolescent[MeSH]). Further, we set a number of search limits: (i) samples aged $\geq$13 and $\leq$18 years; (ii) human species; (iii) studies published between 1st January 2000 and 5th June 2012. This search led to the identification of 413 studies that were screened against a number of selection criteria. Inclusion criteria were: (i) use of sMRI, DTI and fMRI techniques; (ii) samples using substances on a regular basis (as defined by each study protocol). Exclusion criteria were: (i) samples with any major psychopathologies and any neurological disorders; (ii) samples with pre-natal but not current exposure to substances; (iii) non-empirical studies (e.g., review articles, case studies).

The application of these criteria resulted in the identification of 27 empirical studies. A further three studies were obtained by cross-referencing citations reported within the already identified studies [25, 26] leading to a total of 31 studies meeting inclusion criteria. The identified studies were investigated to determine whether: (i) adolescent substance use is associated with adverse neurobiological outcomes; (ii) higher levels of exposure to substances is associated with greater neurobiological harms.

Results

To date, a number of neuroimaging studies (i.e., 31 research investigations) examined the impact of substance use on adolescent samples free of any major psychopathologies. Most of the conducted investigations employed fMRI (n=15), followed by sMRI (n=10) and DTI (n=6). These studies investigated a broad range of substances of use, mostly cannabis (n=11) and alcohol (n=12) alone and combined (n=3). Despite the fact inhalants and ecstasy are also frequently reported to be among the first drugs used by adolescents [4], very few studies have investigated the adverse neurobiological outcomes of adolescent inhalant (n=2), ecstasy (n=1), and polysubstance (i.e., cannabis and tobacco (n=1) or inhalants (n=2)) misuse. See Table 6.1 for a summary of the number of conducted studies by imaging technique and specific substance of use.

Most of the studies examined SUD adolescent samples composed exclusively or mostly by males, with only limited studies investigating samples composed by an equal male to female ratio (these data will be illustrated in 6.1, 6.2 and 6.3 of the following sections). Sample size was relatively low across studies and ranged from 6 [27]–36 [28] participants in both SUD and HC adolescents. Most SUD adolescent samples initiated substance misuse at a mean age of 15 years, while onset of substance misuse ranged between 12 years [29] and 17 years [30, 31] (i.e., mean values).

While all of the reviewed investigations examined regular substance users, research investigating alcohol misuse examined samples that were composed of either alcohol dependent adolescents (1 study [32]), binge-drinkers alone (four studies [29, 33–35]) or in combination (ten studies [28, 36–44]), as well as individuals at high risk of alcohol misuse [32] or with a large range of alcohol exposure (two studies, [38, 45]). Also, while most of the conducted studies examined SUD adolescents who were free of any comorbid psychopathologies, some of the examined samples (two studies [37, 43] were composed of individuals who were diagnosed with psychopathologies that are typically associated with adolescent substance misuse (e.g., ADHD, externalizing disorders [37, 43]). Across the reviewed studies, SUD adolescents were compared to non-substance using controls (HC) with limited exposure to substances, the level of which was defined by each study protocol. A small number of studies ($n=5$) also performed comparisons between different SUDs such as inhalant, alcohol and cannabis use alone and in combination (e.g., DTI [25, 26, 33, 44]. All of the examined SUD groups were matched with control groups for a number of demographic (e.g., age, gender), psychosocial (e.g., SES, education years), neurocognitive (e.g., IQ and measures of learning) and other variables (e.g., handedness). Imaging findings from the literature to date are reviewed by employed imaging technique (i.e., sMRI, DTI, fMRI) in the following sections.

Structural Magnetic Resonance Imaging A total of 10 investigations examined structural brain alterations in adolescent SUD groups by utilising sMRI (see Table 6.2), a widely used imaging technique for both clinical practice and scientific research. Indeed, sMRI provides a measure of structural properties of the brain (i.e., areas, volumes, folding patterns) and allows for the investigation of global and local morphological changes of brain tissue [9]. For instance, sMRI can measure and quantify group differences in the structure of specific brain regions (e.g., hippocampal volume), in cortical thickness and folding (e.g., prefrontal cortex (PFC)), as well as in the shape of specific structures (e.g., corpus callosum) [46].

Altered brain morphology is often ascribed to chronic substance misuse and numerous studies have identified pathological structural brain abnormalities in SUD adolescents, the majority of which were exposed to either cannabis or alcohol (i.e., only one study investigated inhalant users). Brain abnormalities were most consistent for investigations of the PFC, followed by the cerebellum, hippocampus and corpus callosum. Indeed, all of the four studies that examined PFC morphology in

Table 6.1 Number of adolescent studies by type of imaging technique and SUD

Imaging Technique	SUD						
	Cannabis	Alcohol	Inhalant	Ecstasy	Cannabis and Alcohol	Cannabis and Tobacco	Inhalant and Cannabis
sMRI	4	5	–	–	–	–	1
DTI	2	2	1	–	1	–	–
fMRI	5	5	–	1	2	2	–

Table 6.2 Structural MRI studies of SUD adolescents by substances of use

Author (years)	Substance of use	Sample N (Female)		Age (years)		Matching	Level of exposure	Examined areas	Results
		SUD	HC	SUD	HC				
Takagi (2011)	*Inhalant*	14	9	17	19	SUD groups: gender, IQ, education and tobacco/alcohol/cannabis use SUD and HC groups: age, gender, education	Duration 2.5 years	corpus callosum	Inhalant <HC for callosal area and thickness (most markedly in the genu) Inhalant vs. HC had more curved callosal shape
	Cannabis	11		19					
Churchwell (2010)		18(2)	18(6)	18	17	–	Duration 2 years, 1353 use episodes	PFC	CB <HC for OFC than HC (right) CB = HC for OFC (left and total)
Medina (2010)		16(4)	16(6)	18	18	Age, gender, reading and IQ	Duration 3 years, 476 use episodes. Abstinence: 3.5 months	Cerebellum	CB > HC for inferior posterior cerebellar lobules (VIII–X) and vermis volume
Medina (2009)		16(4)	16(6)	18	18			PFC	CB = HC for PFC, PFC/ICV CB female> HC male for PFC CB female HC male for PFC
Medina (2007)		16(4)	16(5)	18	18		Duration 3 years, 476 use episodes. Abstinence: 28 days	Hippocampus	CB = HC for hippocampus
Medina (2008)	*Alcohol*	14(5)	17(7)	17	17	Age, gender, SES, familial SUD, reading, IQ and ethnicity	Duration 3 years, 151 drink episodes, Abstinence 5 days	PFC	AUD < HC for PFC (anterior ventral). Female AUD < HC for PFC (total and WM). Male AUD < HC for PFC (total and WM)

Table 6.2 (continued)

Author (years)	Substance of use	Sample N (Female)		Age (years)		Matching	Level of exposure	Examined areas	Results
		SUD	HC	SUD	HC				
Venkatasu-bramanian (2007)		20(0)	20(0)	16	16	Age, gender and handedness	Less than once/mo	Corpus callosum	At risk AUD < HC for corpus callosum (total, genu, isthmus)
Nagel (2005)		14(5)	17(7)	17	17	Age, gender, education and memory	Duration 2 years, 138 drink episodes	Hippocampus	AUD <HC for hippocampus (left)
De Bellis (2005)		14(6)	24(12)	17	17	Age, gender, IQ, handed-ness, SES	Duration 1 years, 7.3 drinks/occasion	Thalamus, pons/brain-stem, cerebel-lum, PFC	AUD <HC for PFC (total and WM) Male AUD HC for cerebellum
De Bellis (2000)		12(7)	24(14)	17	17	Age, gender, handedness, SES	Duration 2 years	Hippo-campus, amygdala, ventricles, corpus cal-losum, total GM and WM	AUD < HC for hippocampus

SUD substance use disorder, *HC* healthy controls, *N* number, *IQ* intelligence quotient, *OFC* orbial-frontal cortex, *PFC* prefrontal cortex, *ICV* intra-cranial volume, *SES* socio-economic status, *WM* white matter, *AUD* alcohol use disorder, *GM* grey matter. = equal, > greater than, < smaller than

SUD groups found alterations in both cannabis [47, 48] and alcohol users [37, 41]. Further, both of the studies examining the cerebellum consistently demonstrated alterations in both alcohol and cannabis users [37, 49]. These data suggest that PFC and cerebellar regions are particularly affected by adolescent SUD.

Most of the studies that examined hippocampal and corpus callosum morphology also found abnormalities. Hippocampal alterations (i.e., reductions) were found in two studies of alcohol users [42, 43] but not in adolescent cannabis users [50], suggesting a substance-dependent impact on this region in adolescents. Callosal abnormalties were found most prominently in adolescents with comorbid inhalant and cannabis use [26]. Alterations of callosal morphology were also observed in adolescents at risk of alcohol dependence [32], but not in those with alcohol use disorders [43]. Taken together, these findings suggest that adolescent SUD exert a detrimental impact on specific brain regions such as the PFC, cerebellum, hippocampus and corpus callosum. This is in line with previous evidence showing that most of these brain areas, amongst a variety of other examined brain regions, are selectively impaired in SUD adolescents [37].

Morphological abnormalities reported in SUD adolescents may be ascribed to patterns of exposure to substances of abuse. Indeed, measures of prefrontal volumes have been reported to significantly and negatively correlate with measures of alcohol consumption (e.g., drinks per session) [37], recent cannabis use and also alcohol dependence [48]. Similarly, hippocampal abnormalities have been reported to be associated with cannabis dependence [50] and duration of chronic alcohol use [43]. These findings suggest that increasing levels of substance exposure may lead to brain morphological abnormalities that have been observed in SUD adolescents.

Further, earlier age of substance misuse onset may exert a particularly detrimental impact on brain structure in SUD adolescents [43, 47]. Indeed, PFC and hippocampal volumes have been found to be positively associated with age of onset of cannabis and alcohol use, respectively [43, 47]. These results are in line with findings obtained in adult SUD cohorts (i.e., cannabis users), where earlier onset of use was associated with regional brain abnormalities [7]. Thus, earlier onset of SUD appears to be related to brain morphological alterations that are observable in adolescence and that also persist throughout adulthood. Recent findings are however (partially) in contrast with this evidence. Indeed, Cheetham et al. [51] demonstrated that altered PFC, but not hippocampal, morphology predicts SUD onset (i.e., cannabis). Given the paucity of the conducted studies, whether regional brain abnormalities predispose to SUD onset, rather than being an adverse outcome associated with SUD, is yet to be elucidated. Further, evidence on lack of association between patterns of substance exposure (i.e., age of onset, dosage and frequency of use) and regional brain measurements [47] raises questions on the nature of the reported morphological alterations.

Morphological brain abnormalities reported in SUD adolescents have also been suggested to mediate impaired neurocognition and psychopathology symptoms in these groups. Indeed, prefrontal and cerebellar abnormalities have been found to be negatively associated with measures of impulsivity [47] and executive functions [48, 52] in SUD participants. Alteration in these neurocognitive domains, which is associated with impaired decision making (e.g., impulsivity, emotional dysregulation) and a dysregulated ability to reflect on future outcomes, have been related to impairments of PFC functioning [53]; and may be associated in abnormal connectivity between PFC and cerebellar regions that supports high-order cognitive functions [38]. Further, there is preliminary evidence for morphological brain abnormalities (i.e., callosal genu and isthmus) to be associated with symptoms of psychopathologies/disturbed behaviors that are often associated with adolescent SUD (i.e., ADHD, externalizing and internalizing symptoms) [32, 50]. It is important to note that psychopathology symptoms are also associated with impaired neurocognition [54]; and it may be that adolescent SUD may exert an adverse impact on brain regions that mediate the onset of such adverse outcomes. The low number of conducted studies, and the inconsistency between their results (i.e., several studies found no evidence for an association between brain measures and behavioural outcomes [47, 49], prevents, however, determination of the stability and the causality of these associations (i.e., whether psychopathology and impaired neurocognition are a negative outcome or predispose to adolescent SUD).

Summary of sMRI Findings

In summary, these findings suggest that adolescent SUD exert a detrimental impact on PFC, cerebellar, hippocampal and callosal brain regions, most consistently in alcohol and cannabis users. However, there is preliminary evidence that early inhalant misuse is also associated with significant neurobiological harms. Higher chronicity of substance use (i.e., longer duration, higher dosage, substance dependence) may be associated with more persistent abnormalities in SUD adolescents. However, the causality of the association between brain morphology and substance use is yet to be fully elucidated (i.e., it is unclear whether regional brain abnormalities predispose or constitute an adverse outcome of substance use). Findings provided by other imaging modalities may provide further insight on the nature of the association between adolescent SUD and adverse neurobiological outcomes.

Diffusion Imaging: Diffusion Tensor Imaging (DTI) A number of studies (n=6) have examined white matter (WM) integrity in SUD adolescents by employing DTI (see Table 6.3), an imaging technique that provides high-resolution measures/characterisation of the integrity and structural architecture of brain WM fiber tracts [55]. Specifically, DTI measures the freedom with which water molecules can move within WM [56]. In healthy white matter for example, the movement of water molecules within the fibre tracts is restricted by the structural integrity of the axon; in contrast the movement of water molecules may be more diffuse in pathological WM (i.e., with lower integrity) [56]. The two most commonly used DTI measures of WM integrity are mean diffusivity (MD), which quantifies the magnitude of water diffusion (i.e., structural integrity); and fractional anisotropy (FA), which quantifies directionality and white matter tract coherence [57]. In cases of reduced WM integrity (i.e., reduced boundaries to water diffusion), MD is increased and FA is decreased.

Table 6.3 DTI studies of SUD adolescents by substances of use

Author (years)	Substance of use	Sample N (female)		Age (years)		Matching	Level of exposure	Examined areas	Results
		SUD	HC	SUD	HC				
Yücel (2010)	*Inhalant, Cannabis*	11(6)	8(6)	18	20	SUD groups: cannabis and alcohol exposure. SUD and HC: age and gender	Duration: 3 years (inhalant) 4 years (cannabis). Daily or almost daily use for > 1years	Whole brain WM	Inhalant < HC for FA in fibers adjacent to hippocampus and corpus callosum (splenium) CB < HC for FA in same fibers adjacent to the hippocampus Inhalant <CB for FA in corpus callosum
	Cannabis	11(4)		19					
Ashtari (2009)		14(0)	14(0)	19	19	Age, gender, SES, ethnicity and Handedness	Duration 6 years, daily use for at least 1 years, 6 joints/day	Whole brain WM	CB < HC for FA in internal capsule, middle and superior temporal gyrus and arcuate fasciculus
Jacobus (2009)	*Cannabis + Alcohol*	14(2)	14(3)	18	17	Age, gender, IQ	546 life smoking episodes, 153 life drink episodes	Whole brain WM	AUD < HC for FA in corona radiata, longitudinal and fronto-occipital fasciculus. AUD < CB + AUD for FA in corona radiata, inferior fronto-occipital and superior longitudinal fasciculus, middle cerebellar peduncle
	Alcohol	14(2)		18			55 drinking occasions		
McQueeny (2009)		14(2)	14(2)	18	18	Age, gender, education, verbal IQ, ethnicity and SES	55 life drinking episodes, Abst 3 days	Whole brain WM	AUD <HC for FA in corpus callosum, superior longitudinal fasciculus, corona radiate, internal/external capsule, projection fibers from commissural, limbic, brainstem, cortical
De Bellis (2008)		32(7)	28(11)	17	16	Gender and SES	Duration 2 years, 13 drink/week	Corpus callosum	AUD >HC for FA rostral area and isthmus. AUD <HC for MD in isthmus. AUD female < HC female for FA of posterior midbody

Table 6.3 (continued)

Author (years)	Substance of use	Sample N (female)		Age (years)		Matching	Level of exposure	Examined areas	Results
		SUD	HC	SUD	HC				
Bava (2009)	*Alcohol and Cannabis*	36(10)	36(10)	18	18	Age and SES	196 life drinks and 551 life cannabis uses episodes	Whole brain WM	CB + AUD > HC for FA in superior longitudinal fasciculus, postcentral gyrus, crus cerebri, inferior frontal and temporal WM tracts CB + AUD < HC for FA in the occipital, internal capsule, and superior longitudinal fasciculus CB + AUD > HC for MD in the occipital lobe CB + AUD < HC for MD in inferior longitudinal fasciculus

SUD substance use disorder, *HC* healthy controls, *N* number, *FA* fractional anisotropy, *CB* cannabis users, *SES* socio-economic status, *WM* white matter, *IQ* intelligence quotient, *AUD* alcohol use disorder, *MD* mean diffusivity

= equal, > greater than, < smaller than

While a paucity of DTI studies in SUD adolescents has been conducted to date, this research has reported consistent evidence for abnormal WM integrity across all the examined substances of abuse (e.g., cannabis, alcohol and inhalant). For instance, abnormalities were reported in callosal WM fibers across inhalant, alcohol and comorbid alcohol and cannabis users [28, 58, 59]. Alterations in WM tracts of temporo-limbic regions (e.g., fiber tracts adjacent to the hippocampus) were also reported in adolescents with comorbid substance use (i.e., inhalant and cannabis) as well as binge drinkers [58, 60]. WM abnormalities were observed in fiber tracts involved in mediating emotional and neurocognitive processes, the alteration of which is associated to adolescent SUDs. For instance, WM integrity disruption has been reported in fronto-temporal WM fiber tracts, which are ascribed to the regulation of emotional processes, in cannabis and alcohol users (i.e., both alone and in combination) [44, 60, 61]. Thus, alteration of this path may mediate impairments in emotional regulation that have been observed in SUDs. Other impairments of WM fiber tracts have been observed between frontal and other cortices (e.g., sensory, striatal parietal and temporo-parietal cortices) in both cannabis and alcohol users [28, 36, 44], and between cerebellar and other regions (i.e., corticopontal/parietal) [28, 59]. As these paths are involved in mediating higher-order cognitive functions, such findings are consistent with cognitive deficits (e.g., slowed processing speed and memory retrieval) previously reported among SUD adolescents that have been ascribed to WM alterations [62].

While diffuse WM abnormalities have been reported across a range of SUDs, comorbid alcohol and cannabis use was associated with particularly detrimental effects on WM fibre tracts. Indeed, heavy comorbid alcohol and cannabis users exhibited the most marked and widespread alterations in WM integrity [28], compared to WM abnormalities observed in milder users (i.e., lower cumulative exposure to cannabis and alcohol [44]) and in users who consumed only either cannabis, alcohol or inhalants [35, 36, 58, 59, 61].

Notably, the reviewed DTI findings support the notion that increasing levels of exposure may result in greater WM abnormalities. Indeed, there are preliminary reports showing that higher levels of exposure to substances (i.e., alcohol and cannabis [35, 44] and early initiation of use (i.e., inhalant) are associated with lower frontal WM integrity. This evidence is in line with results emerging from sMRI studies of SUD adolescents; and suggests that more chronic, earlier substance use is related to persistent neurobiological harms.

Summary of DTI Findings

Together, DTI results indicate that adolescent SUD, in particular comorbid alcohol and cannabis use, is associated with diffuse WM abnormalities that have previously been ascribed to an arrest or delay in the myelination process [61, 63]. WM alterations affected a variety of brain regions (i.e., callosal and limbic, as well as frontal and sensory, temporal and parietal) mediating developmental processes which occur during adolescence (e.g., emotional regulation and high-order cognitive processes). Preliminary findings also suggest, in line with sMRI results, that chronicity of substance use is related to the reported WM abnormalities. However, the paucity of the reported findings prevents from drawing firm conclusions in this regard.

Functional Magnetic Resonance Imaging Fifteen studies examined the impact of adolescent SUD on the brain by employing fMRI, an imaging technique that indexes changes in brain hemodynamics as indirect markers of brain activity [9]. fMRI attempts to dynamically characterise how the brain functions in real time, which contrasts with the static examination of brain structure that is the focus of sMRI and DTI techniques. Studies of adolescent substance use have focused on characterising brain activity when participants are performing experimental paradigms in the MRI scanner. This 'task driven' approach assumes that task-performance is accompanied by alterations in hemodynamic function. This reflects changes in metabolic demands in localized brain areas due to task-induced neuronal processing [64]. This neuronal processing is assumed to induce changes in the requirements of oxygenated blood to the brain area. The most prominent fMRI measure of hemodynamic function is the blood oxygenation level dependent (BOLD) signal, which is a ratio of oxygenated to deoxygenated blood across brain areas [65]. Magnitude changes in BOLD activation across brain regions is used as the major interpretive variable in fMRI analysis. This has been the focus of fMRI studies of adolescent substance misuse to date (see Table 6.4). Given the limited number of substances examined, the fMRI literature in SUD adolescents will be discussed with respect to the cognitive functions under investigation with the aim of examining consistencies in findings across the examined substances.

Response to Drug-Related Cues

The earliest identified fMRI study of adolescent substance use provides a classical example of how BOLD activation can provide insights into (possible) functional differences in how information is processed in SUD adolescents. Specifically, Tapert and colleagues [39] examined differences in BOLD activation between adolescents with an alcohol use disorder and controls when presented with alcohol-related pictorial advertisements. Presentation of drug-related cues such as advertisements has long been implicated as a potent motivation towards drug seeking in both animal [66] and human [67] studies. Consistent with the adult substance use literature [68, 69], adolescents with alcohol use problems were found to have greater BOLD activation in brain areas consistently linked to reward and craving responses when viewing alcohol-related advertisements compared to neutral pictures [39]. These regions were broadly distributed and included the ventral anterior cingulate and subcallosal areas, in addition to prefrontal, orbital, and limbic regions which together provide evidence for adolescents with alcohol problems being conditioned to have hyper-activity in reward-related brain regions when exposed to alcohol related stimuli, possibly due to an increased salience of these cue [39]. The consistency of this pattern of activation with adult samples suggests that changes in neural processing associated with alcohol cues are present even in early stages of problematic alcohol use. It is unclear however whether these may be present prior to the onset of problematic drinking (which would characterise the majority of adolescent drinkers). However, these results provide interesting insights into alcohol cue induced increases in reward-processing in adolescents with alcohol use problems and suggests that alcohol advertisements may be a potent source of relapse motivation in these individuals [39].

Working Memory

Aside from examining how images of alcohol and alcohol use are processed in adolescents with alcohol use, the most prominent cognitive function that has been examined in adolescent substance users is working memory (i.e., 11 out of 15 studies, see Table 6.4). Working memory is a core executive function that is associated with the ability to attend, maintain, and manipulate information in ready awareness to enable goal directed functioning [70]. Working memory functioning has been one of the most consistent impairments in SUD adolescents for a range of substances such as cannabis, alcohol, and inhalants. Neuroimaging studies have implicated the dorsolateral PFC as being integral in working memory. Still, specific working memory processes may recruit other regions across the brain (e.g., superior parietal or temporal regions) that are required for performance of tasks-specific functions (e.g. spatial verses visual working memory tasks) [71].

Early pilot studies examining working memory in SUD adolescents focused on the hippocampus [30, 72]. Whilst these pilot studies were limited by sample size and perhaps reflecting the still developing fMRI methodology of the time by focusing solely on the hippocampus, both cannabis using [72] and MDMA (ecstasy) using [72] groups failed to show deactivation in the hippocampus when engaged in an auditory working memory task. Emerging evidence suggests that the hippocampus is likely to be important when performing difficult working memory tasks [73] and inhibition of hippocampal neurons has been linked to engaging mnemonic memory processing that may assist with working memory functioning [72].

More recent fMRI studies have examined spatial working memory in adolescent samples of alcohol [38, 40], cannabis [34, 74, 75] and alcohol + cannabis users [29] The spatial working memory task required participants to make a button press response whenever recognizing that an abstract line drawing was presented in a previously presented location. This task therefore engages classically working memory processing whereby participants were required to attend to and maintain task relevant information (e.g. the visualization of the drawing, information about its prior placement) to enable accurate performance. Despite having no behavioural differences in performance of the working memory task, these studies consistently identified different patterns of brain activation between substance misusing groups and controls. This included increased activation in parietal cortical areas relative to controls when performing the spatial working memory task in alcohol [38, 40] and cannabis users [74]. Additionally, these studies have identified reduced activity in cerebellar regions frontal regions [29, 38, 40, 74] although there is also evidence for increased prefrontal activation [29, 40] dependant on the area investigated. Finally, these differences in brain activation when performing spatial working memory tasks may be sex specific with problematic alcohol using females having greater divergence from typical brain activity when performing the spatial working memory task compared to males [40]. Other studies have found similar findings across other areas of the brain. For example, Padula, [75] found an interaction between performance and brain response with "adequate" spatial working memory performance being mediated by differences in activation in temporal regions (superior temporal gyrus, parahippocampal gyrus), anterior cingulate and thalamus in cannabis using adolescents [75].

Finally, in a study examining adolescents who met criteria for both alcohol and cannabis abuse or dependence (CB+AUD), the CB+AUD group were found to have more dorsolateral prefrontal activation, anterior cingulate deactivation, and less activation in left inferior frontal and superior temporal regions compared to controls. Similarly, CB+AUD participants showed significantly more medial frontal deactivation as well as less right inferior frontal and bilateral temporal activation compared to AUD participants [29]. The authors hypothesized that the overall pattern of results in CB+AUD participants suggests that certain task-related regions (e.g., dorsolateral prefrontal cortex) become deficient as a result of substance misuse and other ancillary regions may become active to compensate, resulting in a less efficiently organized neural network. Compared to the CB + AUD group, the AUD participants did not demonstrate the same neural abnormalities, despite similar levels of alcohol consumption, which may indicate that combined cannabis and alcohol may have a unique influence on brain functioning.

Taken together, the differential activation across substance using groups, despite equivalent behavioural performance to controls suggests possible compensatory mechanisms to maintain adequate task performance. This notion converges with evidence from Schweinsberg (2010) who found that recent cannabis users (<7 days abstinence) had greater brain activity in frontal areas compared to abstinent users >27 days abstinence) however these brain regions were not found to be activated when control participants performed the same task. The entirety of this evidence suggests that adolescent substance users may have atypical working memory processing and recruit of brain regions not typically associated with task performance. It is difficult to determine whether these effects are a cause or consequence of substance use. For example, a separate study examining visual working memory in healthy adolescents with a range of substance using patterns found that activity in frontal, cerebellar and parahippocampal regions during the task was correlated with participants' self reported response to the effects of alcohol (e.g. use of alcohol is characterised by strong subjectively rated effects) rather than the amount of alcohol they had consumed. This result suggests that working memory differences in brain activation may be mediated by pre-existing neural differences in the sensitivity to the effects of alcohol. Regardless of this, prospective longitudinal studies may provide a more accurate picture of how substance use may affect working memory functioning in adolescents.

Verbal Learning

Verbal learning is another cognitive domain examined using fMRI in adolescent substance users. For example, Schweinsburg et al. (2010) found that "binge" drinkers had increased activation in frontal and parietal regions when performing a verbal paired associates memory task but reduced activation compared to controls in the hippocampus. Given that the binge drinking group had reduced learning on the task, this evidence suggests that the alcohol using group were utilising suboptimal brain areas for performing the task. Schweinsburg and colleagues (2011) extended this evidence by also investigating cannabis users and binge drinking cannabis users in addition to groups of binge drinkers, and control participants. When performing a verbal paired associates task there were no differences in performance between the groups; however binge drinkers showed higher BOLD response during novel encoding than HC in a range of fronto-parietal regions (see Table 6.4). Overall, both cannabis and binge drinking participants appear to show altered fMRI responses relative to control participants although comorbid alcohol and cannabis exposure may have interactive effects that uniquely alter brain activity (Schweinsburg et al. 2011). In other words, alcohol and cannabis may have synergistic effects that alter brain functioning to a greater extent compared to alcohol or cannabis in isolation.

Inhibitory Control

In addition to verbal learning, one study has examined inhibitory control by employing a Go/No Go task in 28 days-abstinent cannabis users [76]. Despite intact task performance, cannabis users showed higher brain activity during inhibition trials in a range of fronto-medial-parietal and visual cortices [76]. These data suggest that adolescent cannabis users had to place greater physiological demand on their brain (i.e., higher oxygen consumption) to perform at the same level as HC [76]

Summary of fMRI Results

The application of fMRI has proven useful to investigate the neural substrates of cognitive impairments in SUD adolescents (i.e., predominantly cannabis and alcohol using samples) paradigms that engage working memory ($n=9$) and verbal and spatial memory functioning ($n=3$), the impairment of which has been demonstrated

Table 6.4 Functional MRI studies of SUD adolescents by substances of use

Author (years)	Substance of use	Sample N (female)		Age (years)		Matching	Level of exposure	fMRI task	Results
		SUD	HC	SUD	HC				
Becker (2010)	*Cannabis*	EO: 26(7)	–	21	–	Gender and education	Age Onset: EO 13, LO 17 Duration: 4years; Dosage (grams): EO 695, LO 486	Verbal working memory	CB = HC for task performance. EO > LO for superior parietal lobe (left)
		LO: 17(2)		25					
Schweinsburg (2010)A		Cur: 13(4)	18(7)	18	17	Age and gender	Duration: 2 years 342 life use episodes	Spatial working memory	Curr. > Abst. CB and HC for RT speed. Curr. CB Abst. CB for activity in cingulate, medial frontal, superior/middle frontal gyri. Curr. CB > Abst. CB for activity in precentral gyrus
		Abst: 13(4)		18			Duration: 3years. 515 life episodes. Abst: 27–60 days		
Schweinsburg (2008)		Abst: 15(4)	17(5)	18	18	Age and gender	Duration: 3years 481 life use episodes Abstinence: 2 months		CB = HC for task performance. CB HC for activity in dorsolateral prefrontal and occipital cortices. CB > HC for activity in posterior parietal cortex
Padula (2007)		17(3)	17(5)	18	18	Age, gender and IQ	477 life use episodes Abstinence:18 days		CB = HC for task performance. CB > HC for activity in basal ganglia, precuneus, postcentral gyrus and superior parietal lobule

Table 6.4 (continued)

Author (years)	Substance of use	Sample N (female)		Age (years)		Matching	Level of exposure	fMRI task	Results
		SUD	HC	SUD	HC				
Tapert (2007)		16(4)	17(5)	18	18	Age, gender and IQ	Duration: 3years 476 life use episodes. Abstinence:28 days	Inhibition	CB = HC for task performance. *Inhibition trials*: CB > HC for activity in dorsolateral prefrontal, medial frontal, inferior and superior parietal and occipital cortices. *Go trials*: CB > HC for activity in prefrontal, insular, and parietal cortices
Jacobsen (2007)		20(15)	–	17	–	Age, gender and education	847 life cannabis use episodes. 12 daily cigarettes (CB and TOB)	Verbal working memory	Task performance (delayed verbal stimuli recall) was affected by nicotine withdrawal in CB but not TOB. During high verbal working memory load nicotine withdrawal affected CB only for increased activity in posterior cortices and disrupted frontoparietal connectivity
	Tobacco	25(18)		17					
Jacobsen (2004)	*Cannabis*	7(4)	7(5)	17	17	Age, gender and education	282 life cannabis use occasions. Daily cigarettes: 12 (cannabis and tobacco users)	Working memory	CB and TOB = HC for task performance. CB < TOB and HC for hippocampal deactivation
	Tobacco	7(4)		17					

Table 6.4 (continued)

Author (years)	Substance of use	Sample N (female)		Age (years)		Matching	Level of exposure	fMRI task	Results
		SUD	HC	SUD	HC				
Schweinsburg (2011)	*Cannabis*	8(4)	22(6)	18	18	Age, IQ and mood	Duration alcohol: 3 years (AUD), 4 years (CB + AUD). Duration CB: 3 years (CB & CB + AUD). Life drink episodes: 50 (AUD), 210 (CB + AUD). Life CB episodes: 426 (CB), 517 (CB + AUD)		CB and CB + AUD = HC for task performance. AUD showed decreased inferior frontal and increased dorsal frontal/parietal activity. SUD > HC for activity in bilateral frontal regions
	Cannabis and alcohol	28(5)		18				Verbal learning	
	Alcohol	16(3)		18					
Schweinsburg (2005)	*Cannabis and alcohol*	15(5)	19(8)	17	17	Age and gender	Duration: CB 3 years, alcohol 2 years. 310 life CB use episodes (CB). Alcohol use episodes: 135 (AUD) and 129 (CB + AUD)	Spatial working memory	CB and CB + AUD = HC for task performance. CB + AUD < HC for activity in inferior frontal and temporal regions. CB + AUD > HC for activity in other prefrontal regions. CB + AUD > AUD for activity in inferior frontal and temporal activation. CB + AUD < AUD for medial frontal activity
	Alcohol	15(5)		17					

Table 6.4 (continued)

Author (years)	Sub-stance of use	Sample N (female)		Age (years)		Matching	Level of exposure	fMRI task	Results
		SUD	HC	SUD	HC				
Schweins-burg (2010)$_B$		12(2)	12(4)	18	18	Age, gender, IQ and mood	Duration 3 years. 55 life drinking episodes	Verbal learning	AUD < HC for N of recalled words. Novel encoding brain activity: AUD > HC for superior frontal and parietal cortex. AUD < HC for occipital cortex. AUD vs HC did not activate hippocampus
Caldwell (2005)		18(7)	21(9)	17	16	Age, SES and reading scores	Duration 2 years, 155 life use episodes	Spatial working memory and vigilance	AUD > HC for speed of words recall. AUD > HC for activity in superior frontal, superior temporal, cingulate and fusiform gyri. AUD < HC for activity in inferior and middle frontal, precentral gyri, insula, preuneus and cerebellum
Tapert (2004)$_A$		35(22)	–	17	–	–	Alcohol use episodes: 10 (n=3), 10–50 (n=11), 51–100 (n=10), 100(n=10)	Spatial working memory	Brain activity in anterior cingulate, cerebellum and parahippocampus predicted level of alcohol use

Table 6.4 (continued)

Author (years)	Substance of use	Sample N (female)		Age (years)		Matching	Level of exposure	fMRI task	Results
		SUD	HC	SUD	HC				
Tapert $(2004)_B$		15(5)	19(8)	17	17	Age, gender and education	30 life drink episodes	Spatial working memory and motor tasks	AUD = HC for task performance. AUD > HC for activity in parietal cortex. AUD < HC for activity in precentral gyrus and cerebellum
Tapert (2003)	–	15(6)	15(6)	17	16	Gender, SES, education and IQ	50 drinks/month	Alcohol-related cues	AUD = HC for task performance. AUD > HC in left anterior, limbic and visual areas
Jacobsen $(2004)_B$	*Ecstasy*	6(4)	6(4)	17	17	Age, gender, SES, education, cigarette and cannabis exposure	Duration 1 years. 10 life use episodes	Working memory and vigilance	MDMA vs HC did not deactivate hippocampus (left) during high verbal working memory load

SUD substance use disorder, *HC* healthy controls, *N* number, *EO* early cannabis use onset, *LO* late cannabis use onset, *Curr.* current, *Abst.* abstinent, *CB* cannabis users, RT reaction time, *TOB* tobacco users, *AUD* alcohol use disorder, *IQ* intelligence quotient, *SES* socio-economic status in SUD adolescents [77, 78]. Whilst it is difficult to compare studies that examine different cognitive functions, there are consistent differences in brain activity in substance using groups compared to controls particularly, in fronto-parietal areas.

Summary of Findings Across Modalities

First, sMRI has provided evidence for abnormal morphology of brain regions such as the PFC and the cerebellum, while abnormalities in callosal and hippocampal regions were less consistently reported. Second, DTI investigations revealed the presence of widespread abnormalities in WM fibers connecting a variety of areas, including connections between frontal, parietal, temporal and cerebellar regions that are ascribed to the regulation of high-order cognitive processes and emotional states.

Overall, abnormalities were observed across imaging modalities in networks ascribed to the regulation of emotions (i.e., fronto-temporal regions), reward (i.e., fronto-striatal regions) and executive functions (fronto-parietal and fronto-cerebellar regions). There is preliminary evidence that increasing chronicity of substance use (e.g., earlier onset and prolonged use, higher quantity and frequency of use, substance dependence) is related to more pronounced neurobiological abnormalities. Further, altered neurocognition and psychopathologies appeared to be associated with neurobiological alterations in SUD adolescents across the reviewed imaging modalities.

Implications Investigation of the impact of adolescent SUD on regional brain morphology, structural and functional connectivity (i.e., as assessed by sMRI, DRI and fMRI, respectively) is key to elucidate whether and to what degree substance misuse detrimentally affects adolescent neurobiological development that is subject to ongoing remodeling. The findings emerging from this review suggest that structural and functional brain alterations observed in adolescent SUDs may disrupt neurodevelopment. Importantly, neural abnormalities observed in SUD adolescents may persist throughout adulthood. Indeed, investigations of adult SUD cohorts report alterations in regional brain morphology, as well as in structural and functional connectivity that resemble those observed in adolescent SUDs. For instance, structural brain abnormalities have been observed in medial-temporal, cerebellar and prefrontal regions in adult cannabis users [79–82]. Further, widespread alterations in WM and brain function have been observed in prefrontal, cerebellar and limbic areas of adult alcohol [83, 84] and cannabis users [82]. Thus, adolescent SUDs, by altering normal trajectories of neurodevelopment, may lead to neurobiological abnormalities that persist until later in life.

Potential Mechanisms Mediating Neurobiological Abnormalities in SUD Adolescents There is a small but growing literature from both animal and human studies suggesting that adolescent substance misuse disrupts normal neurodevelopmental processes and can induce greater effects on neural plasticity when compared to adult users [85]. The majority of studies report neurobiological harms, however the mechanisms through which these drugs of abuse interfere with normal development are unclear, however several have been proposed. The following sections will illustrate a number of mechanisms that may mediate the adverse impact of adolescent SUD on neurobiological outcomes.

Cannabis

The adverse neurobiological outcomes associated with cannabis have been ascribed to its main psychoactive compound, delta-9-tetrahydrocannabinol (THC). Animal studies have consistently demonstrated that THC exerts detrimental effects on both grey and white matter, and may also disrupt functional activity that relies on the integrity of brain regions. Regular THC exposure has been shown to have neurotoxic effects on neurons and to cause reduction of neuronal bodies, synapses and dendrites [86, 87]. This mechanism has been suggested to mediate the observed brain morphological alterations that have been found in SUD adolescents. The adverse impact of THC on WM fiber tracts may rely on a different mechanism. Indeed, THC has been shown to collect in myelin [88] and CB-1 receptors have been identified in neuronal substrates that are precursors to myelin (e.g., microglia, oligodendrocyte precursor cells) and are essential for healthy neurodevelopment [63, 89]. Prolonged exposure to THC may therefore be associated with CB-1 receptor downregulation (i.e., reduced number and functionality) and result in diminished oligodendrocyte function [63]. Accordingly, regular THC exposure is associated with decreased expression of myelin-related genes [90]. It is hypothesized that the initiation of cannabis use, particularly during adolescence, may alter the trajectory of normal cognitive and white matter development through this mechanism.

Alcohol

It has been hypothesized that intermittent alcohol consumption may induce excitotoxic neuronal damage through an increase in aberrant synaptic activation of NMDA receptors associated with brief episodes of alcohol withdrawal; which can produce marked cellular injury (e.g., hippocampal pyramidal neurons) [91, 92]. Alternatively, alcohol has been shown to decrease neural progenitor cell proliferation and survival in the hippocampus and forebrain of rats during adolescence [93], thus it has been hypothesized adolescent alcohol misuse may disrupt normal neurogenesis (e.g., in the hippocampus) during adolescence and produce cognitive and neurodevelopmental abnormalities.

Importantly, the adverse neurobiological effects of alcohol exposure appear to be enhanced with comorbid cannabis exposure. Indeed, combined alcohol and cannabis use has been demonstrated to exert pronounced neurotoxic effects on brain neurons by triggering apoptotic processes [94].

Inhalants

Inhalants are known to be neurotoxic and readily absorbed into lipid-rich tissues (e.g., the CNS). However, identifying the specific mechanisms through which inhalants interfere with CNS development is complicated because of the heterogeneity of chemicals frequently contained in commonly abused inhalants (e.g., benzene, xylene), but toluene is among the most common chemical contained in frequently abused inhalants and it is hypothesized to be among the most toxic. Animal studies suggest toluene can affect CNS integrity through numerous mechanisms, such as modulating the synthesis and re-uptake of glutamate causing neurotoxicity or activating microglia and secreting proinflamatory cytokines (i.e., neurotoxic immune response) [95].

Overall, adolescent SUD is associated with pronounced neurobiological harms across substances of use (i.e., cannabis, alcohol and inhalants). The nature of the adverse neurobiological outcomes in adolescent SUDs has been mostly ascribed to neurotoxic effects that substances of use, alone and in combination, exert on the morphology, functional and structural connectivity of the brain. To date, the mechanisms through which substance misuse may disrupt normal adolescent neurodevelopment are based on animal models and remain unclear for humans. Indeed, there have been very limited attempts to model how adolescent SUD detrimentally affects neurobiological outcomes in humans. This limitation prevents understanding the mechanisms that drive neurobiological abnormalities reported in SUD adolescents. However, the few imaging studies conducted to date on adolescent substance misuse suggest that the chronic and early initiation of substance misuse leads to neurodevelopmental harms that could persist throughout adulthood.

Limitations A number of methodological limitations of the literature to date limit characterisation of the impact of SUDs on neurobiological outcomes in adolescents. First, only a paucity of studies have investigated the neurobiological outcomes in SUD adolescents, and most of these investigations examined cohorts with a small sample size. These factors hinder our ability to generalize the existing findings to the general adolescent SUD population and to gain sufficient power for detecting more subtle, substance-related brain abnormalities in these cohorts.

Second, most of the conducted studies focused on samples that were composed mostly or exclusively by males. This prevents generalization of the findings to date to cohorts of female SUD adolescents, which are also significantly affected by SUD. Further, this limitation hinders our understanding of gender-specific abnormalities in SUD adolescents; and is an outstanding issue in light of the major influence that sex hormones exert on adolescent brain development. Future studies examining large cohorts with a balanced ratio of males and females are pivotal to further our understanding of this issue.

Third, comorbid psychiatric disorders may have exerted a confounding impact on the adverse neurobiological outcomes of adolescent SUD (i.e., by exerting an independent or an interactive effect). While the present review attempted to minimize comorbid SUD and mental health disorders, a number of studies examined cohorts that were composed of participants with diagnosable or subthreshold psychopathologies that are typically associated with SUD (e.g., ADHD, internalizing/externalizing disorders) [37, 43]. Therefore, it was not possible to disentangle whether the reported neurobiological abnormalities are associated with substance use rather than to mental health disorders (or their interaction). However, given the high comorbidity between SUD and psychopathologies (e.g., anxiety, depression) during adolescence, the reported results may be an accurate reflection of the adverse neurobiological outcomes in the general population of SUD adolescents.

Further, the literature to date has examined a limited number of substances of abuse. Most of the conducted studies examined cannabis and alcohol using cohorts, with a minority of investigations on inhalant, ecstasy and tobacco users, and no study assessing the impact of other substances (e.g., opiates, cocaine, hallucinogens) on the adolescent brain. Thus, there is limited knowledge on the potential harmful effects of substances other than cannabis and alcohol.

Finally, the vast majority of studies investigating adolescent substance misuse are cross-sectional, making it difficult to identify cause and effect relationships. For example, it is possible that reduced grey and white matter volume is a neurobiological risk factor for developing substance use disorders. Longitudinal studies examining samples that are well matched for key variables (e.g., age of regular use, cumulative dose, duration of use) are required to further elucidate the neurodevelopmental impact of adolescent substance misuse.

Clinical Applications for Neuroimaging in Adolescent Substance Misuse

The clinical use of neuroimaging techniques for diagnosis, treatment, and prevention in substance misuse as a whole is still in its infancy. Currently, the vast majority of neuroimaging studies focus on experimental applications for neuroimaging (e.g., identifying potential pathological effects) rather than a clinical tool used for the treatment of substance misuse. However, there is significant potential for neuroimaging to play a role in the treatment of adolescents with addiction disorders.

Cognitive and structural brain abnormalities have been associated with the early initiation of numerous drugs of abuse. For example, significant, diffuse cognitive and neurobiological abnormalities are related to misuse in adults [96], however the progression of these abnormalities from adolescence to adulthood remains unclear. As we develop a greater understanding of the cognitive and neurobiological abnormalities associated with inhalant misuse, structural neuroimaging techniques (e.g., DTI) would be ideal to identify the severity of abuse, chart the progression if abuse continues, and identify candidates for focused clinical interventions.

Similarly, fMRI could be an exceptionally useful clinical tool. Recent fMRI research has examined the relationship between symptom severity and activation in numerous brain networks (e.g., anterior cingulate cortex, parrahippocampay gyrus, striatum) [97]. For example, Hong and colleagues [98] examined cingulate cortex functional circuits in nicotine-dependent individuals using fMRI and found that the severity of nicotine addiction was associated with the functional connectivity strength of dorsal anterior cingulate cortex (dACC) and striatal circuits (i.e., greater dependence was associated with weaker functional activity). The authors hypothesize that the development of new therapeutic techniques aimed at enhancing the dACC-striatum functional network may be effective for nicotine addiction treatment. In other words, therapies that properly engage these neural systems may lower the susceptibility to relapse and improve treatment outcomes for drug users [97].

Conclusions

While significant advances have been made in identifying the harmful neurobiological effects associated with adolescent substance misuse, it remains unclear how drugs of abuse affect cognitive and neurodevelopment during adolescence. There remain many questions in understanding the relationship between drug exposure, neurobiological development, and cognition (e.g., defining the differences in the effects of drug exposure during different stages of adolescent development). However, neuroimaging studies are ideally suited for examining the developmental effects of substance misuse. They are minimally invasive and can yield a diverse range of information about the structural and functional effects of substance misuse over time. Significant clinical applications for MRI techniques are currently being explored and will likely be a useful clinical tool for the treatment of addiction in young people.

References

1. Lubman D, et al. Intervening early to reduce developmentally harmful substance use among youth populations. Med J Aust. 2007;187(7):S22–5.
2. Toumbourou JW, et al. Interventions to reduce harm associated with adolescent substance use. Lancet. 2007;369:1391–401.
3. Johnston L, et al. Monitoring the future: national survey results on drug use. Institute for Social Research, Ann Arbor: The University of Michigan; 1975–2010, 2011.
4. Galduroz J, et al. Household survey on drug abuse in Brazil: study involving the 107 major cities of the country-2001. Addict Behav. 2005;30:545–56.
5. Forster L, Tannhauser M, Barros H. Drug use among street children in southern Brazil. Drug Alcohol Depend. 1996;43:57–62.
6. Musto DF. Drug abuse research in historical perspective. In: Committee on Opportunities in Drug Abuse Research, Editor. Pathways of addiction: opportunities in drug abuse research. 1996. p. 284–94.
7. Wilson W, et al. Brain morphological changes and early marijuana use. A MRI and PET study. J Addict Dis. 2000;19(1):1–22.
8. Pope H, et al. Early-Onset cannabis use and cognitive deficits: what is the nature of the association. Drug Alcohol Depend. 2003;69:303–10.

9. Yang Y, et al., Editors. Structural and functional neuroimaging methods: applications to substance abuse and addiction. West Sussex: Wiley-Blackwell; 2011.
10. Steinberg L, Morris A. Adolescent development. Annu Rev Psychol. 2001;52, 83–110.
11. Spear LP. The adolescent brain and age-related behavioral manifestations. Neurosci Biobehav Rev. 2000;24:417–63.
12. Giedd J. The teen brain: insights from neuroimaging. J Adolesc Health. 2008;42:335–43.
13. Paus T. Mapping brain maturation and cognitive development during adolescence. Trends Cognit Sci. 2005;9(2):60–8.
14. Steinberg L. Cognitive and affective development in adolescence. Trends Cognit Sci. 2005;9(2):69–74.
15. Somerville LH, Jones RM, Casey BJ. A time of change: behavioral and neural correlates of adolescent sensitivity to appetitive and aversive environmental cues. Brain Cognit. 2010;72(1):124–33.
16. Chambers RA, Taylor JR, Potenza MN. Developmental neurocircuitry of motivation in adolescence: a critical period of addiction vulnerability. Am J Psychiatry. 2003;160(6):1041–52.
17. Niendam TA, et al. Meta-analytic evidence for a superordinate cognitive control network subserving diverse executive functions. Cognit, Affect Behav Neurosci. 2012;12(2):241–68.
18. Spear LP. Rewards, aversions and affect in adolescence: emerging convergences across laboratory animal and human data. Dev Cognit Neurosci. 2011;1(14):392–400.
19. Berridge CW. Noradrenergic modulation of arousal. Brain Res Rev. 2008;58(1):1–17.
20. Lubman D, Yücel M. Adolescent substance use and the developing brain. Dev Med Child Neurol. 2008;50:76–80.
21. Luna B, Sweeney JA. The emergence of collaborative brain function: fMRI studies of the development of response inhibition. Ann N Y Acad Sci. 2004;1021:296–309.
22. Bowen S, et al. The last decade of solvent research in animal models of abuse: mechanistic and behavioral studies. Neurotoxicol Teratol. 2006;28:636–47.
23. Slotkin T. Nicotine and the adolescent brain: insights from an animal model. Neurotoxicol Teratol. 2002;24:369–84.
24. Smith R. Animal models of periadolescent substance use. Neurotoxicol Teratol. 2003;25(3):291–301.
25. Yücel M, et al. White-matter abnormalities in adolescents with long-term inhalant and cannabis use: a diffusion magnetic resonance imaging study. J Psychiatry Neurosci: JPN. 2010;35(6):409–12.
26. Takagi M, et al. Corpus callosum size and shape alterations in adolescent inhalant users. Addict Biol. 2013;18(5):851–4.
27. Jacobsen L, et al. Preliminary evidence of hippocampal dysfunction in adolescent MDMA ("ecstasy") users: possible relationship to neurotoxic effects. Psychopharmacology. 2004;173(3–4):383–90.
28. Bava S, et al. Altered white matter microstructure in adolescent substance users. Psychiatry Res. 2009;173(3):228–37.
29. Schweinsburg A, et al. fMRI response to spatial working memory in adolescents with comorbid marijuana and alcohol use disorders. Drug Alcohol Depend. 2005;79(2):201–10.
30. Jacobsen LK, et al. Impact of cannabis use on brain function in adolescents. Ann N Y Acad Sci. 2004;1021:384–90.
31. Becker B, et al. Altered parahippocampal functioning in cannabis users is related to the frequency of use. Psychopharmacology. 2010;209(4):361–74.
32. Venkatasubramanian G, et al. Corpus callosum abnormalities associated with greater externalizing behaviors in subjects at high risk for alcohol dependence. Psychiatry Res. 2007;156(3):209–15.
33. Schweinsburg A, et al. Neural correlates of verbal learning in adolescent alcohol and marijuana users. Addiction. 2011;106:564–73.
34. Schweinsburg A, et al. The influence of recency of use on fMRI response during spatial working memory in adolescent marijuana users. J Psychoactive Drugs. 2010;42(3):401–12.

35. McQueeny T, et al. Altered white matter integrity in adolescent binge drinkers. Alcohol: Clin Exp Res. 2009;33(7):1278–85.
36. De Bellis M, et al. Diffusion tensor measures of the corpus callosum in adolescents with adolescent onset alcohol use disorders. Alcohol: Clin Exp Res. 2008;32(3):395–404.
37. De Bellis M, et al. Prefrontal cortex, thalamus, and cerebellar volumes in adolescents and young adults with adolescent-onset alcohol use disorders and comorbid mental disorders. Alcohol: Clin Exp Res. 2005;29(9):1590–600.
38. Tapert S, et al. Blood oxygen level dependent response and spatial working memory in adolescents with alcohol use disorders. Alcohol: Clin Exp Res. 2004;28(10):1577–86.
39. Tapert SF, et al. Neural response to alcohol stimuli in adolescents with alcohol use disorder. Arch Gen Psychiatry. 2003;60(7):727–35.
40. Caldwell LC, et al. Gender and adolescent alcohol use disorders on BOLD (blood oxygen level dependent) response to spatial working memory. Alcohol Alcohol. 2005;40(3):194–200.
41. Medina KL, et al. Prefrontal cortex volumes in adolescents with alcohol use disorders: unique gender effects. Alcohol: Clin Exp Res. 2008;32(3):386–94.
42. Nagel B, et al. Reduced hippocampal volume among adolescents with alcohol use disorders without psychiatric comorbidity. Psychiatry Res: Neuroimaging. 2005;139:181–90.
43. De Bellis M, et al. Hippocampal volume in adolescent-onset alcohol use disorders. Am J Psychiatry. 2000;157:737–44.
44. Jacobus J, et al. White matter integrity in adolescents with histories of marijuana use and binge drinking. Neurotoxicol Teratol. 2009;31(6):349–55.
45. Tapert, SF, et al. Level of response to alcohol and brain response during visual working memory. J stud Alcohol. 2004;65(6):692–700.
46. Walterfang M, et al. Corpus callosum size and shape in first-episode affective and schizophrenia-spectrum psychosis. Psychiatry Res. 2009;173(1):77–82.
47. Churchwell JC, Lopez-Larson M, Yurgelun-Todd DA. Altered frontal cortical volume and decision making in adolescent cannabis users. Front Psychol. 2010;1(225):1–8.
48. Medina KL, et al. Prefrontal cortex morphometry in abstinent adolescent marijuana users: subtle gender effects. Addict Biol. 2009;14(4):457–68.
49. Medina KL, Nagel BJ, Tapert SF. Abnormal cerebellar morphometry in abstinent adolescent marijuana users. Psychiatry Res. 2010;182(2):152–9.
50. Medina KL, et al. Depressive symptoms in adolescents: associations with white matter volume and marijuana use. J Child Psychol Psychiatry. 2007;48(6):592–600.
51. Cheetham A, et al. Orbitofrontal volumes in early adolescence predict initiation of cannabis use: a 4-year longitudinal and prospective study. Biol Psychiatry. 2012;71:684–92.
52. Medina KL, Nagel BJ, Tapert SF. Abnormal cerebellar morphometry in abstinent adolescent marijuana users. Psychiatry Res. 2010;182(2):152–9.
53. Verdejo-Garcia A, et al. The differential relationship between cocaine use and marijuana use on decision making performance over repeat testing with the Iowa Gambling Task. Drug Alcohol Depend. 2007;90(1):2–11.
54. Yücel M, et al. Understanding drug addiction: a neuropsychological perspective. Aust N Z J Psychiatry. 2007;41(12):957–68.
55. Schmithorst V, et al. Cognitive functions correlate with white matter architecture in a normal pediatric population: a diffusion tensor MRI study. Hum Brain Mapp. 2005;26(2):139–47.
56. Pierpaoli C, et al. Diffusion tensor MR imaging of the human brain. Radiology. 1996;201(3):637–48.
57. Arnone D, et al. Corpus callosum damage in heavy marijuana use: preliminary evidence from diffusion tensor tractography and tract-based spatial statistics. NeuroImage. 2008;41(3):1067–74.
58. Yucel M, et al. White-matter abnormalities in adolescents with long-term inhalant and cannabis use: a diffusion magnetic resonance imaging study. J Psychiatry Neurosci: JPN. 2010;35(6):409–12.
59. McQueeny T, et al. Altered white matter integrity in adolescent binge drinkers. Alcohol: Clin Exp Res. 2009;33(7):1278–85.

60. McQueeny T, et al. Gender effects on amygdala morphometry in adolescent marijuana users. Behav Brain Res. 2011;224(1):128–34.
61. Ashtari M, et al. Diffusion abnormalities in adolescents and young adults with a history of heavy cannabis use. J Psychiatr Res. 2009;43(3):189–204.
62. YücelM,etal.Toluenemisuseandlong-termharms:asystematicreviewoftheneuropsychological and neuroimaging literature. Neurosci Biobehav Rev. 2008;32:910–26.
63. Molina-Holgado E, et al. Cannabinoids promote oligodendrocyte progenitor survival: involvement of cannabinoid receptors and phospatidylinositol-3 kinase/Akt signaling. J Neurosci. 2002; 22(22):9742–53.
64. Logothetis NK. What we can do and what we cannot do with fMRI. Nature. 2008;453(7197):869–78.
65. Heeger DJ, Ress D. What does fMRI tell us about neuronal activity? Nature Rev: Neurosci. 2002;3(2):142–51.
66. Fuchs RA, et al. Relapse to drug seeking following prolonged abstinence: the role of environmental stimuli. Drug Discov Today: Dis Models. 2008;5(4):251–58.
67. O'Brien CP, et al. Classical conditioning in drug-dependent humans. Ann N Y Acad Sci. 1992;654:400–15.
68. Schacht JP, Anton RF, Myrick H. Functional neuroimaging studies of alcohol cue reactivity: a quantitative meta-analysis and systematic review. Addict Biol. 2012;Epub ahead of print.
69. Sinha R, Li CS. Imaging stress- and cue-induced drug and alcohol craving: association association with relapse and clinical implications. Drug Alcohol Rev. 2007;26(1):25–31.
70. Baddeley A. Working memory: theoriestheories, models, and controversies. Ann Rev Psychol. 2012;63:1–29.
71. Wager TD, Smith EE. Neuroimaging studies of working memory: a meta-analysis. Cognit, Affect Behav Neurosci. 2003;3(4):255–74.
72. Jacobsen L, et al. Preliminary evidence of hippocampal dysfunction in adolescent MDMA ("ecstasy") users: possible relationship to neurotoxic effects. Psychopharmacology (Berl). 2004;173(3–4):383–90.
73. Jeneson A, Squire LR. Working memory, long-term memory, and medial temporal lobe function. Learn Mem. 2012;19(1):15–25.
74. Schweinsburg A, Brown S, Tapert S. The influence of marijuana ise on neurocognitive functioning in adolescents. Curr Drug Abus Rev. 2008;1(1):99–111.
75. Padula C, Schweinsburg A, Tapert S. Spatial working memory performance and fMRI activation interactions in abstinent adolescent marijuana users. Psychol Addict Behav. 2007;21(4):478–87.
76. Tapert S, et al. Functional MRI of inhibitory processing in abstinent adolescent marijuana users. Psychopharmacology. 2007;194:173–83.
77. Bates ME, et al. Short term neuropsychological recovery in substance use disordered clients. Alcohol: Clin Exp Res. 2005;29(3):367–77.
78. Tapert S, et al. Substance use and withdrawal: neuropsychological neuropsychological functioning over 8 years in youth. J Int Neuropsychol Soc. 2002;8:873–83.
79. Cousijn J, et al. Grey matter alterations associated with cannabis use: results results of a VBM study in heavy cannabis users and healthy controls. NeuroImage. 2012;59(4):3845–51.
80. Lorenzetti V, et al. Structural MRI findings in long-term cannabis users: what do we know? Subst Use Misuse. 2010;45(11):1787–808.
81. Jager G, et al. Effects of frequent cannabis use on hippocampal activity during an associative learning memory task. Eur Psychopharmacol. 2007;17:289–97.
82. Lopez-Larson MP, et al. Altered prefrontal and insular cortical thickness in adolescent marijuana users. Behav Brain Res. 2011;220(1):164–72.
83. Pfefferbaum A, Adalsteinsson E, Sullivan E. Supratentorial profile of white matter microstructural integrity in recovering alcoholic men and women. Biol Psychiatry. 2006;59(4):364–72.

84. Pfefferbaum A, Sullivan E. Disruption of brain white matter microstructure by excessive intracellular and extracellular fluid in alcoholism: evidence from diffusion tensor imaging. Neuropsychopharmacology. 2005;30(2):423–32.
85. Lubman D, Yücel M, Hall W. Substances and the adolescent brain: a toxic combination? J Psychopharmacol. 2007.
86. Landfield PW, Caldwallader LB, Vinsant S. Quantitative changes in hippocampal structure following long-term exposure to Δ9-tetrahydrocannabinol: possible mediation by glucocorticoid systems. Brain Res. 1988;443(1–2):47–62.
87. Chan GCK, et al. Hippocampal neurotoxicity of Δ9-Tetrahydrocannabinol. J Neurosci. 1998;18(14):5322–32.
88. Ashton JC, Darlington CL, Smith PF. Co-distribution of the cannabinoid CB1 receptor and the 5-HT transporter in the rat amygdale. Eur J pharmacol. 2006;537(1–3):70–1.
89. Sanchez C, et al. Delta-9 tetrahydrocanabinol induces apoptosis in C6 glioma cells. FEBS Lett. 1998;436:6–10.
90. Davis KL, et al. White matter changes in schizophrenia evidence for myelin-related dysfunction. Arch Gen Psychiatry. 2003;60(5):443–56.
91. Guerri C, Pascual M. Mechanisms involved in the neurotoxic, cognitive, and neurobehavioral effects of alcohol consumption during adolescence. Alcohol. 2010;44:15–26.
92. Hendrickson A, et al. Aberrant synaptic activation of N-methyl-D-aspartate receptors underlies ethanol withdrawal hyperexciteability. J Pharmacol Exp Ther. 2007;321(1):60–72.
93. Crews F, et al. Neurogenesis in adolescent brain in potently inhibited by ethanol. Neuroscience. 2006;137:437–45.
94. Hansen HH, et al. Cannabinoids enhance susceptibility of immature brain to ethanol neurotoxicity. Ann Neurol. 2008;64(1):42–52.
95. Win-Shwe T, Fujimaki H. Neurotoxicity of toluene. Toxicol Lett. 2010;198:93–98.
96. Rosenberg N, et al. Neuropsychologic Impairment impairment and MRI abnormalities associated with chronic solvent abuse. Clin Toxicol. 2002;40(1):21–34.
97. Reske M, Paulus M. The diagnostic and therapeutic potential of neuroimaging in addiction medecine. In: Adinoff B, Stein E, editors. Neuroimaging in addiction. Chichester: Wiley-Blackwell; 2011.
98. Hong L, et al. Association of nicotine addiction and nicotine's actions with separate cingulate cortex functional circuits. Arch Gen Psychiatry. 2009;66(4):431–41.

Michael Takagi The University of Melbourne, Melbourne Neuropsychiatry Centre, Department of Psychiatry. Psychiatry PhD.

George Youssef The University of Melbourne, Melbourne Neuropsychiatry Centre, Department of Psychiatry. The University of Melbourne, Department of Psychology, Psychological Sciences. MPsych(Neuro) PhD.

Valentina Lorenzetti The University of Melbourne, Melbourne Neuropsychiatry Centre, Department of Psychiatry. PhD in Psychiatry.

Chapter 7
Alcohol Abuse in Adolescents: Relevance of Animal Models and Experimental Results in Adolescent Animals

Rosana Camarini and Ricardo M. Pautassi

Introduction

This chapter reviews studies comparing the effects of drugs on adolescents and adults. The disorders associated with drug dependence usually start in adolescence or early adulthood. Among the significant consequences of drug use in adolescence is a higher risk of abuse and dependence in adulthood. The aim of this chapter is to describe the main studies on the behavioral, neuroanatomical, neurochemical, and functional differences between adolescents and adults and to correlate such differences with the higher vulnerability of adolescents to dependence. The hypothesis that adolescents might exhibit an idiosyncratic pattern of reactivity to ethanol that might expose them to higher risk of use/abuse, with alcohol also serving as a gateway for other drugs, is also assessed. The focus of this chapter is alcohol because it is one of the drugs most widely used by adolescents.

Why is it Relevant to Assess Alcohol Use in Adolescence?

Alcohol (or ethanol) use is considered a public health concern worldwide. Some authors found that in the United States, 7 out of 10 adolescents consume alcohol before finishing high school [1]. Such worrisome use of alcohol by adolescents also

R. Camarini
Department of Pharmacology, Institute of Biomedical Sciences, USP, São Paulo, Brazil
e-mail: camarini@icb.usp.br

R. M. Pautassi
National Council of Scientific and Technological Research, Buenos Aires, Argentina
Department of Psychology, National University of Córdoba, Córdoba, Argentina

D. De Micheli et al. (eds.), *Drug Abuse in Adolescence*,
DOI 10.1007/978-3-319-17795-3_7

occurs in Latin America, with the prevalence of use at 50% among individuals aged 12–17 years old in Brazil [2], Argentina [3], and Mexico [4]. A recent study conducted with a birth cohort from Pelotas (Brazil) found that by age 11–12 years old, 20% of the children had already tried alcohol [5]. At the end of the study, alcohol use was found to occur during pregnancy in certain cases, while a study conducted by Alderete et al. [3] detected symptoms of dependence in 12% of the girls and 19% of the boys.

Children who consume alcohol tend to exhibit poorer school performance and higher odds of engaging in risk behaviors, undesired sexual behaviors, conflict with the law, and traffic accidents. Long-lasting contact with alcohol at a young age is associated with long-term consequences. Epidemiological studies showed that alcohol use by youths is a risk factor for alcohol-related problems. In a study assessing young alcohol users, Grant and Dawson [6] found a negative association between age at onset and the prevalence of dependence throughout life. In that large study (sample size: 26,616), approximately 40% of the participants who reported having used alcohol before age 15 years old were diagnosed as dependent at some point in their lives. That proportion was significantly lower (10% and 25%) among the participants who had started drinking at age 17–22 years old, respectively. Those data seem to allow a simple explanation of the negative impact of early exposure to alcohol use: the participants who had started drinking earlier had more time to develop dependence compared to the participants who had started later. However, that possibility was eliminated by DeWit et al. [7]. Those authors measured alcohol abuse and dependence per group of age at onset adjusted by the time elapsed since initial use (in years). Statistical analysis detected a rapid progression from alcohol abuse to dependence among the individuals who had started drinking before age 14 years old but not among the individuals who had started drinking at age 19 years old. Curiously, at least one large epidemiological study found that relative to alcohol use disorders, early onset of drinking is synergic with exposure to stressful situations [8].

Several mechanisms might be involved in the so-called early debut effect, so that exposure to ethanol might alter the normal development of the trajectories of critical brain areas and certain brain transmission systems. Certain brain areas, in particular the prefrontal cortex, which participates in the regulation and planning of behavior, seem to be singularly sensitive to the toxic effects of alcohol. Binge drinking (heavy episodic drinking) reduces adolescents' performance in neuropsychological tasks requiring integrity of the prefrontal cortex. Binge drinking can be defined as the intake of large doses of alcohol in a single episode, attaining a blood alcohol concentration $\geq$80 mg/dl.

Another explanation of the permissive effect related to alcohol abuse based on premature exposure considers social interaction with peers, which is intensified during adolescence. Within that scope of assessment, use of and preference for alcohol might be affected by the perception of reinforcing and intoxication patterns induced by alcohol in peers ("passive social influence") [9]. Our study and others derive the existence of a causal relationship between premature exposure to alcohol and increased odds of abuse and dependence [7, 8, 10]. Based on that conclusion, new policies should be formulated to hinder access to alcohol or to delay its use until early adulthood. Early onset of drinking might also be considered to be a consequence of a preexisting pathology, such as an externalizing behavior disorder arising from the socio-familial environment.

Use of Animal Models to Analyze Drug Use and Risk Factors Associated with Drugs in the Course of Development and Their Effects Along Life

There are certain inconsistencies with regard to the onset of adolescence, particularly when a general definition is attempted with disregard for species. A more conservative definition considers the postnatal developmental stage from age 28 to 42 days old as adolescence in rodents, although animals weaned between ages 21 and 60 days are often also considered to be adolescents. In addition to the logistical simplicity associated with the use of animals in different laboratories, those models allow for substantial quality control, shorter breeding time, hypothesis testing, and precise genetic control, enabling the performance of a wide variety of pharmacological manipulations to investigate the role of neurotransmitters and hormone systems upon the onset and maintenance of drug use. In addition, a large number of behavioral techniques were elaborated and refined to assess motivated and autonomic behaviors and memory deficits due to alcohol use concerning the modulation of those effects by other relevant variables, as well as the ones derived from early environmental exposure to stressful events.

The advantages associated with the use of animal models to describe recent advances focusing on the use of drugs in adolescence are discussed here. The passive social influences (e.g., peer pressure) that might be relevant for motivating drug seeking in adolescents and young adults were mentioned above. Social animal models for the transmission of the preference for alcohol in adolescence employ the "observer-demonstrator" paradigm. According to that paradigm, an animal that had recent access to food (demonstrator) is permitted to interact with another animal considered "naïve" (observer). When the latter is assessed in a test in which it has to choose between two places, it exhibits greater preference for the one with food (target). Hunt et al. [11] promoted interactions between adolescent rats acting as observers or demonstrators, where the latter had been given alcohol (0.0, 1.0, 1.5 or 3.0 g/kg), decaffeinated coffee, or water, to then measure the ethanol intake. Only the observers that had interacted with demonstrators previously given 1.5 g/kg ethanol exhibited increased intake, indicating that social contact with an ethanol-intoxicated peer alters the pattern of ethanol consumption. Fernández-Vidal and Molina [9] tested the preference for the smell of alcohol in adolescents before and after interaction with demonstrators given 1.5 g/kg ethanol 30 min earlier. The observers were placed in a box to explore two compartments, one with alcohol and the other with a vanilla-scented substance. The animals that had interacted with ethanol-intoxicated demonstrators exhibited an increased preference for the smell of alcohol. Those findings suggest that the adolescent observers perceived the ethanol in their demonstrator peers, and such experience might influence their future use of the drug.

In studies conducted among humans, the genetic predisposition for alcohol abuse has been traditionally understood on the basis of a family history positive for alcohol dependence (FH^+). Among other possibilities, FH^+ individuals are more sensitive to the peripheral and reinforcing effects of alcohol, which are usually experienced while the blood alcohol concentration is increasing. A genetic predisposition to alcoholism was successfully mimicked in animal models through the development of lineages of animals selected based on their level of alcohol intake [12]. Some such lineages are HAD, LAD, AA, ANA, P, and NP. In a recent study, Pautassi et al. [13] used animals with (P) and without (NP) alcohol preference bred at Indiana University [12]. The aim of that study was to assess whether a high predisposition to alcohol intake manifested immediately after birth or a previous experience with the drug was needed. After alcohol was offered by means of artificial nipples, the P newborn rats consumed more than twice as much alcohol as the NP ones, which suggests a genetic component in alcohol use. That preference reverted in the second week of life, to reappear at the beginning of adolescence.

Animal models of adolescence are also useful for testing hypotheses arising from epidemiological studies that indicate that the use of alcohol at an early stage, increases the likelihood of serious alcohol-related problems [7] and facilitates its use vis-à-vis stressful events that routinely occur in life [8]. Using an animal model that involved intermittent binge drinking in adolescence, Pascual et al. [14] found significant changes in the mesolimbocortical pathway, affecting the dopaminergic and glutamatergic systems in particular, which are associated with increased alcohol use in adulthood. Hargreaves et al. [15] found that access to beer for 24 h every 3 days (considered to represent an intermittent pattern of access; alcohol concentration 4.44 %) induced the binge drinking pattern in adolescent rats but not in adult rats. As mentioned above, binge drinking is highly prevalent among adolescents and might induce significant neurocognitive deficits.

In addition, preclinical studies provided some hypotheses and emphasized some variables that might somehow be used in epidemiological studies with humans. Increasing numbers of studies are using multivariate models (structural equation modeling, analysis of trajectories, and cluster analysis) to test theoretical models for symptoms related to drug dependence. Those analyses allowed the detection interrelationships and of direct and indirect effects between variables first identified in preclinical studies. The results of those combinations of studies in both animals and humans led to a more robust conception of the etiology of chemical dependence, which is thus currently considered to be a developmental disorder, the causes of which might occur within the first years of contact with the drug (childhood and adolescence) [16].

Differences in the Sedative Effects and Locomotor Deficit Induced by Alcohol and in the Ability to Metabolize it

As has long been known, after being given similar doses, the blood and brain alcohol concentrations (BEC and BrEC) are higher in women than in men. In addition, the women exhibit poorer motor coordination, which is usually explained by the fact that women have less body water and mass compared to men. Similar age-related differences might increase the likelihood that women will exhibit mental problems resulting from drug use. Similarly, adolescents also tend to exhibit greater BEC and BrEC compared to adults after being given similar doses of alcohol. That phenomenon is not specific to adolescence but is the result of the elimination rate, which increases more or less linearly during the course of development (from approximately 4.5 mg/dl/h in newborn rats to 42 mg/dl/h in adult rats) [17]. These differences pose a challenge to epidemiological studies because the age-related differences in the effects of alcohol might be confounded with different plasma ethanol levels. This problem might be overcome in animal studies by using alcohol doses that, albeit different, induce similar BEC and BrEC [18]. As a function of the aims of studies, another possible method to overcome this problem might be to use alcohol doses that induce the same level of alteration with respect to age. Considerable evidence indicates that compared to adults, adolescent rats are relatively insensitive to several effects of alcohol; such insensitivity would operate as a "barrier" hindering the occurrence of other behaviors related to drug intake. In one study, Varlinskaya and Spear [19] gave different doses of alcohol or saline solution (control group) to rats and then let them interact with "sober" peers, i.e., animals that had not consumed alcohol. Alcohol induced slight effects on the adolescents, but the intoxicated adult rats tended to contact their peers much less and to avoid the other animals. The ability of alcohol to induce sleep and psychomotor disorders is significantly lower in adolescent rats than in adult rats. On the contrary, the ethanol-induced cognitive deficits seem to be more intense in adolescents than in adults, particularly relating to tasks that demand hippocampal integrity. In addition, adolescents are more sensitive to alcohol-induced social interaction behaviors. Their idiosyncratic response pattern might increase the likelihood that adolescents will develop mental disorders following drug use.

Differences in the Perception of the Reinforcing Effects of Alcohol

The sedative effects of alcohol and the induced motor deficits are significant enough to require regulation of its use. The apparent insensitivity of adolescents to certain effects allows them to drink in excess (for instance, binge drinking). Alcohol intake is mainly regulated by the balance between its reinforcing and aversive motivational consequences. The reinforcing effects of alcohol increase the odds of seeking and use, while perceiving its aversive effects results in a significant reduction in intake [16].

Few studies have assessed the motivational effects of alcohol with regard to age; a slightly larger number of studies have assessed conditioned taste aversion (CTA) to alcohol. For that purpose, animals are paired between a new flavor and the flavor of alcohol. Several authors induced reduction of conditioning in adolescents compared to adult rats [20], although both age groups expressed a similar level of aversion upon being conditioned to high alcohol doses within the context of another learning task called second-order conditioning (SOC) [10].

In SOC, a discrete stimulus (CS1) is paired with the effects of alcohol. CS1 is thus contingent to a salience (CS2; e.g., a rough texture), and the preference for CS2 is measured. Application of this model allowed a highly relevant finding: only adolescent rats expressed preference conditioning when the alcohol dose was kept low [18]. That finding indicates increased sensitivity to the appetitive effects of alcohol in those animals and is consistent with the fact that adolescents, but not adults, self-administer alcohol doses sufficient to induce tachycardia. Certain authors who conduct research with humans understand those alcohol-induced changes in heartbeats to be based on the appetitive effects of alcohol, while authors who perform studies with mice usually favor ethanol-induced locomotor activity (LMA). Certain studies conducted with mice have found high LMA sensitivity in adolescent compared to adult animals. However, those results have not been reproduced in studies conducted with rats, perhaps because some rodent species are more insensitive to the stimulating effects of alcohol than others. Recently, some authors found that adolescent rats expressed greater LMA following use of high but not of low doses of alcohol (2.5 and 0.5 g/kg, respectively). Perhaps one of the most important findings in that study was that the animals that exhibited a high LMA rate also exhibited high alcohol intake compared to the animals that exhibited a low LMA rate. It is worth noting that although LMA is considered to be equivalent to other classic reinforcement paradigms, such as conditioned place preference, that association is still hypothetical, and therefore, further studies are needed to elucidate the functional meaning of those behaviors. It might be reasonable to consider LMA to be a sub-product of the anxiolytic effects of alcohol and that the latter might act as a negative reinforcer.

To summarize, although there is little evidence, adolescents appear to be more sensitive to the reinforcing effects of alcohol and less sensitive to its aversive effects. In other words, the reinforcing effect of alcohol might be stronger in youths than in adults. Thus, youths seem to express a pattern of response to alcohol that, as a result of the combination of specific sensitivities and insensitivities, might put them at risk of controlled substance use.

Main Neurochemical Findings in Adolescent and Adult Animals in Response to Drugs of Abuse

Accurate comprehension of the motivational mechanisms in adolescents is crucial to understand why so many youths consume alcohol and engage in risk behaviors. The distinct neuroanatomical, neurochemical, molecular, and physiological characteristics of adolescence that distinguish it from other stages in life might account for certain behaviors inherent to that developmental stage. Preclinical studies found age-related variations in sensitivity to the effects of various drugs of use, as follows. Their impulsiveness and greater proneness to risk behaviors make adolescents more vulnerable to dependence on drugs of abuse. The impulsiveness typical of adolescence is characterized by a poor ability to perform critical judgments on rewards, such as drugs of abuse, sex, food, and money. That behavior has been associated with delayed development of the prefrontal cortex. Incomplete maturation of the prefrontal cortex in adolescence limits inhibitory capacity vis-à-vis impulsive behaviors.

What are the typical anatomical and functional differences of the adolescent brain that provide evidence for such behaviors? Throughout adolescence, the neural circuits undergo remodeling and fine adjustment, particularly in the cortical areas (frontal, temporal and parietal cortices) and the hippocampus. That neurodevelopment involves the resizing and reorganization of synapses and increased myelination of axons in the prefrontal cortex. The absolute volume of the prefrontal cortex is reduced in adolescent humans and rats. Maturation allows improvement of the efficacy and speed of the neural mechanisms and contributes to the control of impulsiveness, as well as to the working memory and thinking and to decision-making. The prefrontal cortex plays a relevant role in judgment, decision-making, and control of emotional responses and is responsible the execution and inhibition of motivational behaviors. Other functions are associated with memory consolidation, planning, problem solving skills, and logical thinking.

Studies conducted in humans and animals point to the existence of a primary motivational circuitry encompassing the prefrontal cortex and ventral striatum, which influence the response of motor structures responsible for decision-making [21]. Dysfunctions in that area are associated with impulsive behavior and increased risk of drug use. The mesolimbic dopaminergic system is composed of a set of neurons, the bodies of which are in the mesencephalic ventral tegmental area and that project especially to the nucleus accumbens and prefrontal cortex, in addition to other limbic areas. That system is associated with reward mechanisms, reinforcing the effects of drugs of abuse, and risk attitudes. In addition to the dopaminergic system, glutamatergic afferent projections to dopaminergic neurons in the nucleus accumbens and striatum modulate the synaptic plasticity that is associated with addiction. Both the dopaminergic and glutamatergic systems undergo significant changes throughout development, particularly during adolescence, as the striatum, nucleus accumbens, prefrontal cortex, and hippocampus that undergo significant maturational changes. Significant loss of glutamatergic excitatory synapses that arrive to the prefrontal cortex occurs in humans and non-human primates. In turn, the dopaminergic and serotonergic afferent fibers increase in that area, reaching peaks much higher than in all other stages in life. The connectivity between the prefrontal cortex and subcortical structures continues developing throughout adolescence; this phenomenon is demonstrated by the increased density of prefrontal-accumbens projections, as well as by the density of dopaminergic and glutamatergic innervation of the mesencephalic ventral tegmental area and amygdala. In adolescent rats, the dopamine levels were shown to increase in the striatum, accompanied by lower dopamine release in the nucleus accumbens following pharmacological or environmental stimulation. Reward events might induce sensitization of striatal dopamine release in adolescents compared to adults [22]. Those maturational changes in these neural circuits are associated with cognitive functional alterations, attention, working memory, and response inhibition.

Ontogenetic alterations in receptors are also described in the literature. Several studies have reported peaks in the cortical production of dopaminergic and glutamatergic NMDA (N-methyl-D-aspartate) receptors at the onset of adolescence, which then significantly decline until the onset of adulthood [13, 23]. The peak amount of D1 and D2 receptors in the striatum and nucleus accumbens is 30–45 % higher than in adulthood.

The GABA receptor comprises a heterogeneous set of subunits, where subunit $\alpha1$ is involved in sedative, amnestic, and anticonvulsant actions. That subunit increases in the prefrontal cortex in childhood and adolescence, to then decrease in the transition to adulthood. The expression of subunit $\alpha2$, which is responsible for anxiolytic properties, increases in the frontal cortex during adolescence; that subunit has been associated with the genetics of alcoholism. The neurochemical and neuroanatomical immaturity and greater neuroplasticity of the mesocorticolimbic system, which is a part of the reward system that modulates the motivation for natural and non-natural (e.g., drugs of abuse) reward, might contribute to adolescents' greater vulnerability to dependence.

To summarize, the main neurotransmission systems are not mature at birth but continue developing throughout adolescence, with remodeling being more accentuated in the prefrontal cortex and limbic areas. The maturation of the prefrontal cortex is evidenced by increased neural activity in that area, contrasting with a reduction in the activity of irrelevant brain areas. Compared to adolescents, adults exhibit greater activation of specific areas that play a critical role in response inhibition, such as the inferior frontal gyrus, which shows that the response inhibition capacity improves with age.

In addition, a temporal dissociation has been posited relative to the maturation of the prefrontal cortex and sub-limbic areas. As a result of the earlier development of the limbic system vis-à-vis the prefrontal cortex, the limbic areas are functionally mature, while the prefrontal cortex is immature during adolescence. That is, there is an imbalance between the limbic system (related to pleasure) and the prefrontal cortex (related to response inhibition) in adolescence. By comparison to childhood, when both systems are undergoing development, and adulthood, when both are fully mature, adolescence is characterized by certain behavioral traits, such as impulsiveness and high-risk decisions, that might be correlated with the development of the neurobiological bases that underlie those behaviors. According to that model, under extreme emotional situations, the activity of the already mature limbic system prevails over the inhibitory control of the still immature prefrontal cortex. The impulsiveness typical of adolescence has been associated with the immature ventral prefrontal cortex, while high-risk behavior seems to derive from increased nucleus accumbens activity.

As a function of the structural, synaptic, cell, and neurochemical organization of the brain areas related to the reward circuit in adolescence, the impact of drugs of abuse on the brain functions might be expected to be greater in adolescents than in adults. Animal models demonstrate differential responses to drugs of abuse as a function of age, which most likely derive from the functional and structural changes undergone by the maturing brain compared to the mature brain.

The sensitivity of adolescents to the acute locomotor effects of drugs of abuse is still a subject of controversy in the literature. The results of related studies are controversial, and the effects vary as a function of age, type of drug, and laboratory. Thus, both increased and reduced locomotion were reported, as well as the absence of any effect of nicotine on adolescents. The results relative to alcohol and cocaine are equally inconclusive. Adolescents respond less to the locomotor-stimulating action of amphetamine compared to adults; however, the response to morphine is more accentuated in adolescents. Additionally, the response to behavioral sensitization induced by drugs of abuse is conflicting. The range of sensitization in adolescence and adulthood might depend on the type of drug, dose, length of exposure, and pairing of the drug effects with the environment. Relative to nicotine, some researchers found reduced sensitization in adolescents compared to adults, but others registered a greater effect. Studies that investigated the response to sensitization to alcohol in adolescents found that adolescent rats were less sensitive than adults. Concerning psychostimulants, there are reports that adolescents become more sensitized to the locomotor-stimulating effects of amphetamine than adults. There is not yet a consensus with regard to cocaine in this regard.

Relative to reward effects, several studies agree that nicotine and ethanol are more rewarding for adolescent rats than adults. The conditioned place preference induced by amphetamine and cocaine occurs with lower doses in adolescents compared to the doses required by adults, although the self-administration levels are similar in both. Regarding consumption, adolescent rodents consume more alcohol than adults; although the results are variable, as a rule, the frequencies of self-administration and seeking of nicotine and cocaine, respectively, are also higher in adolescents.

Adolescents are less sensitive than adults to various ethanol effects, such as sedation, locomotor impairment, and withdrawal effects. Such insensitivity to the acute effects of ethanol allows adolescents to consume larger amounts of ethanol over longer periods of time than adults without perceiving its intoxicating effects.

Adolescents seem to be more sensitive to the neurotoxic effects of drugs of abuse compared to adults: for instance, greater predisposition to memory impairment, as evidenced by spatial memory tests. Intermittent use of alcohol in adolescence induces inflammatory damage in the prefrontal cortex and hippocampus, which is associated with long-lasting behavioral and cognitive impairment. The mechanisms underlying those lesions seem to be related to alcohol-induced neuroinflammation. The latter involves activation of the microglia via the stimulation of toll-like receptors (TLR4) and the induction of cytokines and other inflammatory mediators (iNOS, NO, COX-2). Intermittent administration of ethanol to adolescents increases the inflammatory mediator levels and consequently also cell death in the cortex, hippocampus, and cerebellum. Cognitive impairment occurs concomitantly with the increase in the inflammatory mediator levels and can be prevented through the administration of non-steroidal anti-inflammatory drugs.

Ethanol-induced prefrontal cortex injury in adolescence seems to lead not only to cognitive dysfunctions but also to a predisposition to abuse, impulsiveness, and dependence. In tests that assess learning involving the orbitofrontal cortex, which demands non-spatial working memory and reference memory, the performance of rats that consumed cocaine in adolescence was poorer than the performance of rats that were given the drug in adulthood. That finding is particularly relevant because it has been suggested that in drug dependence, the orbitofrontal cortex is activated by extremely attractive motivational stimuli only (such as, e.g., cocaine in the case of dependence). Thus, drug-seeking behavior is maintained at the expense of the normal functioning of the learning processes, particularly when the cognitive demands are high. Such sequestration of the orbitofrontal function might account for the neurocognitive impairment exhibited by rats that self-administer cocaine.

The simultaneous use of various drugs is increasingly common among youths. The combination of alcohol and ecstasy causes severe memory impairment in adolescent rats, as evidenced by impaired performance in the radial arm maze test and reduction of the granule neuron population in the hippocampal dentate gyrus. Molecular and physiological differences between adolescents and adults at least partially explain their differential responses to drugs of abuse and the greater vulnerability of the former to dependence.

As mentioned above, adolescents exhibit risk behaviors and seek novelties, which promotes "heavy drinking" and trying other drugs. Adolescent animals exposed to ethanol are more prone to exhibiting cognitive disorders, including memory and learning problems. Considering the behavioral and physiological differences between adolescents and adults, we describe below the main neurochemical findings related to drugs of abuse.

It is well known that drugs of abuse increase the release of dopamine in the nucleus accumbens. Studies have shown that exposure to drugs such as ethanol and cocaine in adolescence might induce greater dopaminergic sensitization and faster achievement of the dopamine peaks, respectively, compared to adults, which might be a sign of greater expectation for the drug. Differential alterations in the glutamatergic system were also reported. Ethanol is known to inhibit glutamatergic activity; however, the NMDA receptor subunits NR2A and NR2B are particularly sensitive to the effects of alcohol. Intermittent treatment with alcohol was associated with reduced NR2B receptor phosphorylation in the prefrontal cortex, hippocampus, and nucleus accumbens in adolescent rats but not in adults. In addition, the repeated administration of ethanol to adolescent mice was shown to increase the release of glutamate in the nucleus accumbens, while the opposite effect was observed in adults. The induction of immediate response transcription factors, such as c-fos, in response to the administration of drugs of abuse differs between adolescents and adults. Following the administration of low but not of high doses of nicotine, adolescents exhibited greater c-fos expression in the nucleus accumbens "shell". High doses of cocaine induced greater c-fos expression in the striatum of adults compared to adolescents. Repeated administration of ethanol decreased c-fos expression in the prefrontal cortex and nucleus accumbens in adolescent mice but not in adults. The opposite effect was detected in the hippocampus. The regulation of the gene fos protein product, delta FosB, is also age-dependent. Following repeated treatment with cocaine or amphetamine, adolescent mice expressed more delta FosB in the nucleus accumbens and caudate putamen than adults. In adolescent rats exposed to alcohol, the preference for and consumption of alcohol increased in adulthood. Those behavioral effects exhibited association with reduced protein levels of D2 receptors and the phosphorylated 2B subunit of glutamatergic receptor NMDA.

Epigenetics involves modifications in the activation of some genes through the remodeling of chromatin, but not of the basic DNA structure, mediated by histone acetylation or methylation. Recent studies have revealed the relevance of epigenetic changes in the effects of alcohol and cocaine during adolescence. Reduction of histone H3 methylation in the prefrontal cortex following the administration of cocaine to adolescents seems to be associated with long-term behavioral consequences. While repeated treatment with alcohol increased the acetylation of histones H3 and H4 in the prefrontal cortex and nucleus accumbens and reduced the acetylation of those histones in the striatum in adolescent rats, no changes were detected in adults. It was suggested that changes in chromatin remodeling might play a role in the long-lasting behavioral alterations induced by early drug use.

Neurochemical and epigenetic changes were shown to be associated with long-lasting behavioral consequences following the discontinuance of drug administration. Therefore, those neuroadaptations bear serious implications, especially in the case of the developing brain.

References

1. Johnston LD, O'Malley PM, Bachman JG, et al. Monitoring the future national results on adolescent drug use: overview of key findings, 2008. Bethesda: National Institute on Drug Abuse; 2009.
2. Galduroz JC, Carlini EA. Use of alcohol among the inhabitants of the 107 largest cities in Brazil, 2001. Braz J Med Biol Res. 2007;40:367–75.
3. Alderete E, et al. Problemas relacionados con el consumo de alcohol en jóvenes de la provincia de Jujuy, Argentina. Salud Pub Mex. 2008;50(4):300–7.
4. Rojas-Guiot E, et al. Alcohol and drug consumption among students in Pachuca, Hidalgo. Salud Pub Mex. 1999;41(4):297–308.
5. Noal RB, et al. Experimental use of alcohol in early adolescence: the 11-year follow-up of the 1993 Pelotas (Brazil) birth cohort study. Cad Saúde Pub. 2010;26(10):1937–44.
6. Grant BF, Dawson DA. Age at onset of alcohol use and its association with DSM-IV alcohol abuse and dependence: results from the National Longitudinal Alcohol Epidemiologic Survey. J Subst Abuse. 1997;9:103–10.
7. DeWit DJ, et al. Age at first alcohol use: a risk factor for the development of alcohol disorders. Am J Psychiatr. 2000;157:745–50.
8. Dawson D, Grant A, Bridget F, et al. Impact of age at first drink on stress-reactive drinking. Alcohol Clin Exp Res. 2007;31(1):69–77.

9. Fernández-Vidal JM, Molina JC. Socially mediated alcohol preferences in adolescent rats following interactions with an intoxicated peer. Pharmacol Biochem Behav. 2004;79(2): 229–41.
10. Pautassi RM, et al. Ethanol induces second-order aversive conditioning in adolescent and adult rats. Alcohol (NY). 2011;45(1):45–55.
11. Hunt PS, Holloway JL, Scordalakes EM. Social interaction with an intoxicated sibling can result in increased intake of ethanol by periadolescent rats. Develop Psychobiol. 2001;38: 101–9.
12. Bell RL, et al. The alcohol-preferring P rat and animal models of excessive alcohol drinking. Addict Biol. 2006;11(3–4):270–88.
13. Pautassi RM, et al. Ontogeny of ethanol intake in alcohol preferring (P) and alcohol nonpreferring (NP) rats. Develop Psychobiol. 2011;53(3):234–45.
14. Pascual M, et al. Repeated alcohol administration during adolescence causes changes in the mesolimbic dopaminergic and glutamatergic systems and promotes alcohol intake in the adult rat. J Neurochem. 2009;108(4):920–31.
15. Hargreaves GA, Monds L, Gunasekaran N, et al. Intermittent access to beer promotes binge-like drinking in adolescent but not adult Wistar rats. Alcohol. 2009;43(4):305–14.
16. Casey BJ, et al. The adolescent brain. Dev Rev. 2008;28:62–77.
17. Kelly SJ, Bonthius DJ, West JR. Developmental changes in alcohol pharmacokinetics in rats. Alcohol Clin Exp Res. 1987;1:281–6.
18. Pautassi RM, et al. Adolescent but not adult rats exhibit ethanol-mediated appetitive second-order conditioning. Alcohol Clin Exp Res. 2008;32(11):2016–27.
19. Varlinskaya EI, Spear LP. Acute effects of ethanol on social behavior of adolescent and adult rats: role of familiarity of the test situation. Alcohol Clin Exp Res. 2002;26:1502–11.
20. Anderson RI, Varlinskaya EI, Spear LP. Ethanol-induced conditioned taste aversion in male sprague-dawley rats: impact of age and stress. Alcohol Clin Exp Res. 2010;34(12):2106–15.
21. Alfonso-Loeches S, Guerri C. Molecular and behavioral aspects of the actions of alcohol on the adult and developing brain. Crit Rev Clin Lab Sci. 2001;48:19–47.
22. Laviola G, et al. Risk-taking behavior in adolescent mice: psychobiological determinants and early epigenetic influence. Neurosci Biobehav Rev. 2003;27:19–31.
23. Teicher M, Andersen SL, Hostetter Jr JC. Evidence for dopamine receptor pruning between adolescence and adulthood in striatum but not nucleus accumbenus. Dev Brain Res. 1995;89:167–72.

Rosana Camarini BS in Pharmacy and Biochemistry, University of São Paulo (USP). MA and PhD in Psychobiology, Federal University of São Paulo (UNIFESP). Postdoctoral fellowship, University of California, San Francisco USA. Associate Professor, Department of Pharmacology, Institute of Biomedical Sciences, USP. President, Latin American Society for Biomedical Research on Alcoholism.

Ricardo M. Pautassi Adjunct Researcher, National Council of Scientific and Technological Research. Professor, Department of Psychology, National University of Córdoba, Argentina.

Chapter 8
FMRI Studies of the Adolescent Reward System: The Triadic Model Perspective

Rista C. Plate, Jessica M. Richards and Monique Ernst

The opinions and assertions contained in this paper are the private views of the authors and are not construed as official or as reflecting the views of the NIMH or the Department of Health and Human Services.

The goal of this review is to describe maturational changes of the neural systems that could explain the adolescents' propensity for drug abuse. We start with the presentation of a heuristic neural systems model, the triadic model, which could help guide neuroscience research into motivated behaviors. A systematic review of the developmental neuroimaging studies of the reward system provides a basis to begin to delineate the neural mechanisms contributing to vulnerability to substance abuse in adolescents.

This review is organized in two sections. The first section provides descriptive and theoretical constructs of adolescent behavior and underlying neural substrates. The second section covers the literature on neuroimaging findings reported in studies of reward-related paradigms. In the first section, we address the reasons why adolescence is such an important period to understand for neuroscientists, particularly from the perspective of motivated behaviors and reward sensitivity. Then, the description of the specific aspects of motivated behaviors in adolescence leads to the presentation of the triadic model. We particularly emphasize the reward node of this model because of its utmost relevance to substance abuse, which is the focus

M. Ernst (✉)
Section on Development and Affective Neuroscience, National Institute of Mental Health Intramural Research Program, NIMH-Building 15-K, Room 110, MSC-2670, Bethesda, MD 20817-2670, USA
e-mail: ernstm@mail.nih.gov

R. C. Plate
Department of Psychology, University of Wisconsin-Madison, NIMH-Building 15-K, Room 110, MSC-2670, Bethesda, MD 20817-2670, USA
e-mail: rplate@wisc.edu

J. M. Richards
Department of Psychiatry and Human Behavior, Johns Hopkins University School of Medicine, 5510 Nathan Shock Drive, Ste 100, Baltimore, MD 21224, USA

D. De Micheli et al. (eds.), *Drug Abuse in Adolescence*,
DOI 10.1007/978-3-319-17795-3_8

of this book. The second section of this review first details the strategies used in functional neuroimaging studies to compare reward-related processes in youth and adults in particular, and then presents the findings accumulated so far. We conclude by underscoring the questions most in need of research, particularly those aimed at resolving the discrepancies in the literature and clarifying the dynamics that regulate the balance within the triadic model.

Adolescence, Motivated Behavior, and the Triadic Model

Adolescent Behavior

Adolescence marks a distinct transition period with unique characteristics and behaviors. This period represents both a time of opportunity for building the roots of a successful and fulfilling adult life, and a time of vulnerability owing to the adverse consequences of the typical impulsive/risky adolescent behaviors and the unique vulnerability to mental problems. The most commonly recognized behaviors typical of this period include cognitive impulsivity, risk seeking, emotional intensity and lability, and social reorientation [1, 2, 3]. As a result, adolescents are at an increased risk of engaging in hazardous behaviors, such as tobacco and drug use, risky sexual activity, or reckless driving (e.g., [4, 5, 6], which can result in unplanned pregnancy, sexually transmitted disease, bodily harm, or even death [1, 7, 8]. In addition, adolescence is a prime period for the onset of mental illness [9, 10]. Finally, interindividual variability within this stereotypical description of adolescent behaviors is large, and can be traced to hormonal changes, early life experience, genetic make-up, among many other factors. A better understanding of the underlying factors contributing to interindividual variability can be tremendously helpful for identifying targets for future primary and secondary treatment of untoward outcomes of adolescence. We now turn to the description of each of the typical adolescent behaviors.

Cognitive impulsivity, a hallmark of adolescent behavior, reflects the difficulty in waiting for an expected outcome. One way to operationalize this effect is to frame the delay as a cost that is larger for adolescents than adults. Researchers have tested cognitive impulsivity using tasks in which participants can either pursue stimuli with smaller immediate rewards or stimuli with larger, but delayed, rewards [11, 12, 13, 14]. Possible neural mechanisms underlying developmental differences in discounting of delayed rewards include a unique functional pattern of inhibitory control, such that adolescent impulsivity may be affected by the diminished ability to evoke top-down neural control centers [15].

Risk seeking is another typical behavior of adolescents. This behavior manifests in various ways, for example organized activities such as mountain climbing or parachute jumping, or more problematic actions such as drug use or unprotected sex [16, 17]. Outcome uncertainty defines risk, such that risky behaviors entail potential positive outcome and/or potential harm to the individual [18]. Experimental assessment of risk seeking behavior is difficult due to its complexity as well as the challenge of creating real-world situations that not only accurately measure how adolescents would react, but also effectively predict future behaviors. Several studies attempted to create such experimental tasks with some success [19–26]. Indeed, performance on risk-taking paradigms has been shown to correlate with real-world risk behaviors measured with questionnaires [27, 28]. In addition, developmental studies generally indicate that risk seeking peaks in adolescence [21–23, 29–32].

Emotional intensity and lability in adolescence has been noted across centuries. For example, Dahl [1] recounts historic literary examples of adolescent idealism and pursuits of romantic passion to illustrate emotional lability and affective intensity. Adolescents commonly fluctuate between extreme positive and negative emotions, expressing elation in one moment and apathy or frustration the next. Affective lability has not yet been extensively studied; however, the research completed so far shows that emotional lability is particularly severe for adolescents with depression [33]. More research needs to be conducted to define and characterize affective lability and intensity and determine how it can best be assessed in an experimental setting.

Social reorientation from parents to peers constitutes a drastic change in the social landscape of adolescents. Adolescents spend more time with peers than they spend with adults [8, 34]. This shift to a peer-centered social life has considerable impact on motivated behaviors. Adolescents report more positive than negative affect in the presence of peers compared with time alone [33]. In addition, adolescents take more risks within a social than nonsocial context [35], and are acutely sensitive to the views of their peers [36]. While adolescents may act at adult levels of maturity in the absence of peers, peer influence can detrimentally affect an adolescent's ability to think through situations and act responsibly and effectively. Research shows that, when decisions are made in situations that evoke strong emotions ("hot" situations, such as among peers), adolescents take more risks compared to adults. In contrast, adolescents perform similarly to adults in "cold" situations, without heightened emotionality [30].

Brain Development

Substantial neural development accompanies the rise in impulsivity, emotionality, and risk seeking over the course of adolescence. The normative trajectory of adolescent neural development is becoming well-characterized with the advent of MRI technology. Specifically, the adolescent brain undergoes not only structural and functional neural changes, but evidence also suggests that individual brain structures and regions have unique and asynchronous developmental trajectories. Furthermore, these individual trajectories may be orchestrated along distinct timelines, which are critical to the harmonious development of brain function (see Fig. 8.1) [37, 38].

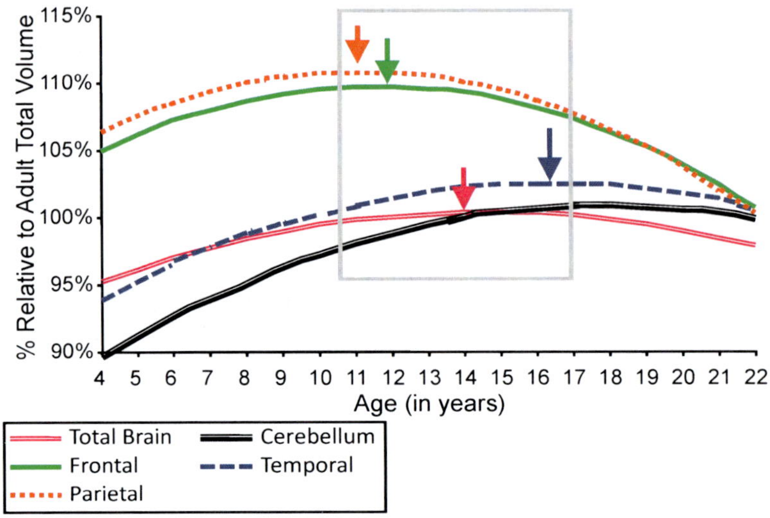

Fig. 8.1 Developmental trajectory of individual brain regions. Distinct brain regions reach adult maturity along variable, chronologically determined, time courses. Deviance in the chronology of the brain systems development may affect adolescents' abilities to successfully recruit and control such systems. Arrows represent points of inflexion, when the curves change direction

At the histological level, substantial cell, dendrite, and synapse proliferation and then elimination proceed with time [39, 40] (see Fig. 8.2). These cellular processes lead to more selective and refined information processing. Together with axonal caliber enlargement, myelination (the formation of an insulated white sheath around the nerve fibers) and increase in axon diameter, contributes to the decrease in gray matter and the increase in white matter with age [41–44]. Myelination speeds up the transmission of information over long distances (e.g., cross-hemispheric projections), and ultimately provides more efficient transmission of information.

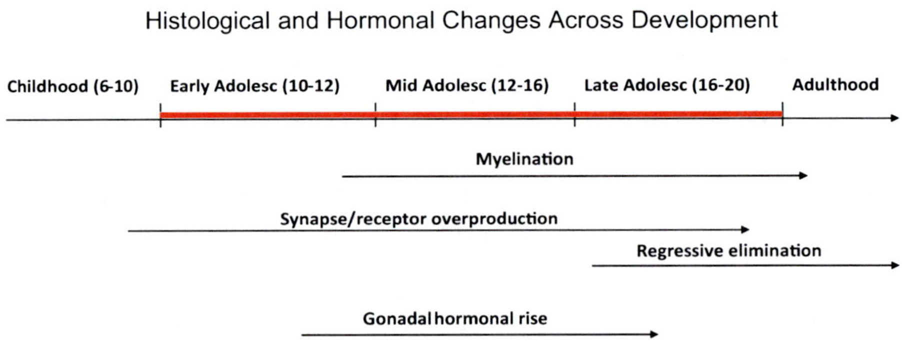

Fig. 8.2 Brain development with age. Several changes occur at the histological and hormonal levels and continue across development. Synapse overproduction begins in mid-to-late childhood and is followed later by regressive elimination in late adolescence. Myelination, associated with the increase of white matter and related decrease in gray matter begins in early adolescence and continues through young adulthood. Gonadal hormonal rise, part of puberty, characterizes this period

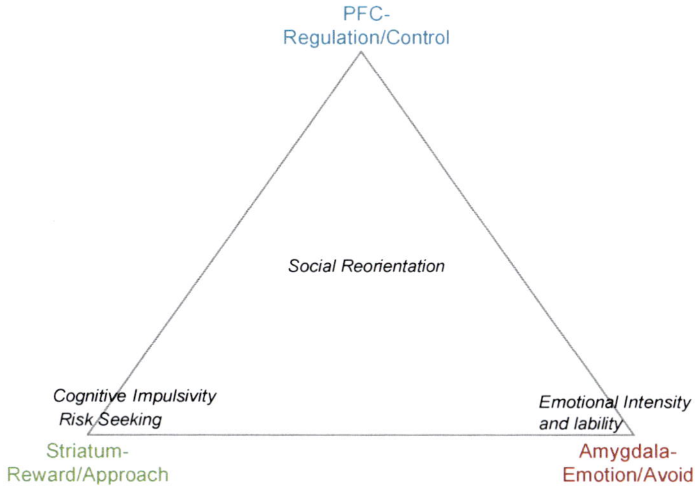

Fig. 8.3 The triadic model. The prefrontal cortex (*PFC*) has a reciprocal relationship with the striatum and amygdala, and the amygdala projects directly to the striatum. Within the triadic model the striatum represents the reward system, and is associated with approach; the amygdala represents the emotion system, particularly responses to aversive (e.g., fearful) stimuli, and plays a significant role in avoidance; and the prefrontal cortex is the regulatory center, which serves to control approach and avoidance behaviors. Of the four behaviors typically observed in adolescence, the striatum is chiefly responsible for risk seeking and cognitive impulsivity; the amygdala for emotional intensity and lability. Social reorientation involves interactions among all three systems

The notion of a predetermined timetable of the progression of various neural changes (e.g., loss of grey matter is last in the superior temporal cortex; [42]) has critical implications. Indeed, important behavioral consequences can ensue following a disruption within this predetermined order. These behavioral consequences could result from the sub-optimal coordination among nodes or modules of behavioral controls, and not necessarily from selective local regional abnormalities. This scenario emphasizes interregional influences (i.e., functional connectivity) and the role of potential imbalances in neural maturation across various brain regions, each of which is implicated in specific behavioral patterns.

The triadic model of the neurobiology of motivated behavior in adolescence is based on such neural systems organization [45]. This model suggests that normative maturation of the neural systems, involved in distinct behavioral processes, occurs along a predetermined order (see Fig. 8.1 [38]). In particular, three networks are proposed to interact and code for patterns of behavioral responses to stimuli. These three networks underlie respectively *reward* with a privileged association with appetitive motivational processes, *emotion* with a privileged association with negative emotion, and *regulatory* function over emotion and motivated behavior (see Fig. 8.3). The specific neural regions emphasized in the model include the striatum given its role in reward, approach, and habitual behaviors [46–48]; the amygdala for its involvement in emotion, threat, and social information processing [49–52];

and the prefrontal cortex (PFC), which modulates affective and cognitive processes [53–57]. These modules refer to reward, emotion, and regulation from the perspective of their dominant function, and to approach, avoidance, and control from the perspective of their behavioral correlates. The next section will describe the triadic model in greater detail.

The Triadic Neural Systems Model

The platform of the three modules (approach/reward, avoidance/emotion, and control/regulation), which compose the triadic model, provides a basis for studying behavioral responses, and more specifically here, the typical adolescent behaviors that were reviewed above, i.e., cognitive impulsivity, risk seeking, emotional intensity and lability, and social reorientation. How these adolescent behaviors can be operationalized using the triadic model template is schematized in Fig. 8.3. *Cognitive impulsivity* and *risk seeking* reflect a hyperactive reward module (serving to approach stimuli or situations) combined with a unique modulation of the emotion-related module (enhanced delay cost, and reduced avoidance of potential negative stimuli or situations, respectively), as well as a hypoactive control region unable to regulate increased reward-seeking. *Emotional intensity and lability* indicate poor regulation of emotional responses, a reflection of the poor capacity of the regulatory module to modulate the emotion network. Finally, *social reorientation* represents a switch in social value, both in terms of magnitude (affective intensity) and quality (switch from familial to peer), and may reflect a re-attribution of positive and negative values to social stimuli [2]. This shift in social orientation is likely to originate from the interaction among all three maturing modules of the triadic model in addition to a reorganization of the social neural circuitry [58, 59].

The triadic neural systems model provides a framework for the basic decomposition of a simple behavioral response (Fig. 8.3): Upon encountering a stimulus, one can respond in two ways, approach or avoid. The direction of the response (approach or avoid) is regulated by the *regulatory module* which assigns a unique weight to the neural systems underlying approach (*reward module*) and avoidance (*emotion module*). These three distributed neural modules are distinct but largely overlapping, with reciprocal functional interconnections [2]. The triad reaches an equilibrium that is specific to a given situation, but that is modulated by both transient and sustained factors. Transient factors include individual mental state (e.g., depressed, stressed), physical state (e.g., drug action), or shift in context (e.g., school vs. home, social vs. non-social). Sustained factors include individual mental traits (e.g., inhibited temperament), maturation level (age; puberty), genetic make-up, past experiences, or gender. These factors are critical as they contribute to the large inter-individual variability in behavioral responses. The present review is concerned mainly with the effect of age on the triadic function. However, this model can be used to examine the effects of disorders on the neural function underlying motivated behavior. Before presenting functional neuroimaging studies that address

Table 8.1 Neural substrates of the triadic model: The anatomy, function, and role of the striatum, amygdala, and prefrontal cortex

Modules of the Triadic Model		
Approach	**Avoidance**	**Regulation**
	Main structures	
Striatum	Amygdala	Dorsolateral PFC
Orbitofrontal cortex	Hippocampus	Ventral PFC
	Insula	Anterior cingulate cortex
	Function	
Appetitive stimuli	Aversive stimuli	Salience detection
Valence/salience value	Valence/salience value	Executive attention
Motivation	Fear responses	Motor control
Motor response	Threat avoidance	Conflict detection
Positive affect	Negative affect	Conflict monitoring
		Conflict resolution inhibition

the effect of age on the triadic model in a reward context, we briefly review the neural substrates of the components of the triad (see Table 8.1; Fig. 8.4).

The approach module refers to the reward-related neural system. This neural system comprises subcortical and cortical structures that are major sites of dopamine

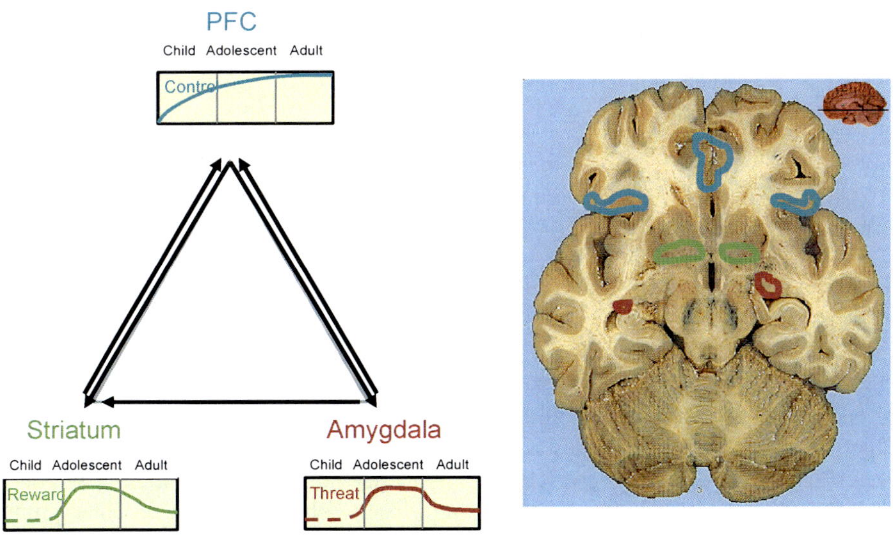

Fig. 8.4 Developmental pattern of neural substrates of the triadic model. The PFC develops monotonically with age, with typical adults possessing the most mature, capable systems. The striatum and its related systems show a hyper-responsive peak in adolescence, as does the amygdala in its response to threat

action, and include primarily the striatum (caudate nucleus, putamen and nucleus accumbens) and the medial and orbital prefrontal cortices [47, 60]. Behaviorally, appetitive motivational processes seem to follow a curvilinear developmental trajectory, whereby reward sensitivity peaks in adolescence [61].

The avoidance module refers to the emotion-related neural system. Although this system is involved in both positive and negative emotions, it is uniquely implicated in threat-related processes (e.g. [50]). This module comprises the amygdala, hippocampus, and insula, which are consistently associated with response to aversive stimuli [52, 62]. Behaviorally, emotion-related processes also seem to follow a curvilinear function, by which emotional responses peak in intensity in adolescence [63– 66].

The control module refers to regulatory processes that modulate subcortical function, (i.e., the approach and the avoidance systems), through "top-down" cognitive control. This module relies on prefrontal cortical structures that carry specialized functions, such as inhibition (right inferior prefrontal cortex) [54, 67, 68], working memory and cognitive salience detection (dorsolateral prefrontal cortex) [69], conflict detection, monitoring and resolution (anterior cingulate cortex) [53, 55, 56]. Behaviorally, control processes mature monotonically with age [39, 70, 71].

The next section reviews functional neuroimaging studies that can elucidate the differential recruitment of the modules of the triadic model across adolescence. We will restrict the review to the tasks that have been used to compare adolescents, adults, and, for some, children, on the properties of the reward system. The reward system is the most relevant system to addiction, which has been characterized as the hijacking of the reward system function by drugs of abuse [72]. Furthermore, according to the triadic model, functional imbalance of the reward module relative to the control module can result in the persistent use (approach) of drugs (reward) despite harmful consequences.

fMRI Task Paradigms to Assess Decision Making and Motivation

Basic Principles Underlying Reward-Related Paradigms

Before reviewing findings informing the functional characteristics of the reward system in adolescents relative to adults, the various paradigms probing specific aspects of this system will be described. Figure 8.5 summarizes the general approach used to study the neural correlates of reward in the fMRI environment, which is based on the decomposition of simple motivated behaviors into elementary functional units of behavior. The basic structure of an fMRI paradigm (task) involves the repetition of trials. A minimum of 20 repetitions of each type of trial is considered necessary to obtain a reliable fMRI signal. This empirical number depends on the type of task, and can be determined a priori by task modeling algorithms. Each trial

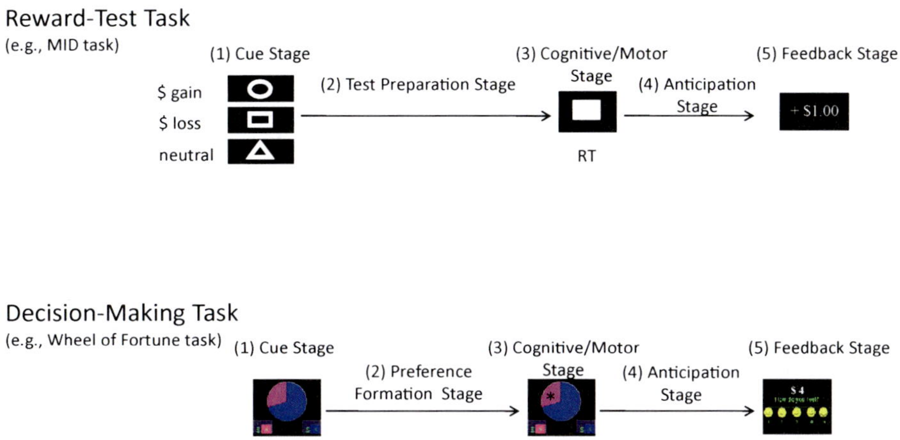

Fig. 8.5 Reward paradigm stages: An appraisal *cue stage* (provides trial information or signals participants to prepare to respond), a *preference/preparation stage* (formation of preference for decision-making paradigms and performance preparation for the reward test-tasks), a *cognitive/motor stage* (execute the selection or the test); an *anticipation stage*; and a reward *feedback stage*

is composed of the basic processes of interest. Most reward-related tasks present time-windows for reward *anticipation* and reward *feedback*. Other processes can be assessed in other stages of the task structure, as described below.

Rewards, across the available paradigms, can vary on multiple factors, for example featuring appetitive stimuli such as pleasant pictures (ice cream, happy faces) or monetary gains. These paradigms can manipulate the frequency, intensity (salience), rewarding quality (valence: positive, negative or neutral stimuli), sensory domain (visual, olfactory, gustatory, tactile), probability of reward receipt, or conditioning level (e.g., primary vs. secondary) of the rewarding stimuli. These rewards can also be preceded by a warning signal (cue), which provides a window for examining reward anticipation. These manipulations are applicable to nearly all types of tasks of reward function.

Three basic task structures have been used to contrast reward-related processes in adolescents relative to adults. The first and simplest task consists of the *passive presentation of rewards* without any active action on the part of the individuals to obtain the rewards. One example of a passive reward presentation paradigm, the slot-machine task [73], has been used to contrast adults and adolescents on the passive receipt of probabilistic gains. We also include here an emotion task, which presents three types of emotionally salient stimuli (positive, negative and neutral) in the form of portraits displaying various emotions [74]. Although used in the context of emotion processing rather than reward/motivation processing per se, this task is a perfect example of a passive viewing paradigm of appetitive (happy expression), aversive (angry or fearful expressions), and neutral (neutral expression) stimuli. One limitation of these passive viewing tasks is the absence of behavioral or subjective measures that can inform the nature of individual differences in neural activation.

The second type of task, *reward test-task,* requires participants to complete an action (test) to obtain the reward, and this action is not directly related to the reward itself (see Fig. 8.5, upper panel). These actions can involve cognitive and/or motor processes that need to be performed correctly to obtain the reward. For example, the test could consist of a memory challenge (e.g., the Pirate's paradigm [75]), a perceptual judgment (e.g., the Cake Gambling task [76]), or a timed motoric action (e.g., the monetary incentive delay (MID) task [77]). Many factors can be manipulated in these paradigms, including the difficulty (accuracy level) and the nature of the processes involved in the test (e.g., cognitive, motor, perceptual). This type of task can also be preceded by a signal (cue stage) informing the nature of the trial to come (e.g., reward vs. loss), or the amount of reward associated with the trial. Overall, reward test-tasks permit researchers to examine the interactions between reward processes and cognitive/motor processes (e.g. [78–80]).

The third type of task is a *reward decision-making task* (see Fig. 8.5, lower panel). Participants are required to select one of several options, each associated with a distinct probability of reward. This type of task is probably the most complex as it involves a number of different behavioral reward-related processes, each providing opportunities for experimental manipulations. The *formation of a preference* guiding the decision-making is the process uniquely associated with this third type of task. The boundary between a test task and a decision-making task can be subtle, and subject to controversy. For example, we place the Cake Gambling task [76] in the category of test-tasks, because there is an uncontroversial correct and incorrect response, the correct response being the largest slice of the cake. This selection, therefore, does not depend on subjects' subjective *preference*, but on a perceptual discrimination of size. However, because this test has been framed as making a decision for the reward underlying the largest color area, this task is usually referred to a "reward decision-making task". Overall, decision-making reward tasks permit researchers to test preference among risky, novel, or delayed options. Finally, a combination of the three types of task permit researchers to probe the effect of agency on reward processes [81]. The studies described below are summarized in Table 8.2 and Table 8.3. Table 8.2 introduces the studies, categorized by type of reward task, and includes the populations studied, paradigm, and stage(s) analyzed. Table 8.3 presents the major developmental findings of the studies, classified by type of reward-task, stage, and activation areas. In addition, we provide the peak activation coordinates, when available, using either the Talairach (Tal) or the standard Montreal Neurological Institute (MNI) coordinates.

Findings of Reward-Related fMRI Studies

This section is organized by type of paradigms (as described above) and stage of the reward process (see Fig. 8.5). As seen in Fig. 8.5, five stages can be distinguished, a *cue stage* (presentation of information about the trial, or signal to get ready, as a simple orienting stimulus), a *preference formation stage* for the decision-making

Table 8.2 Tasks used to probe developmental neural correlates of reward. The studies are arranged according to type of reward task and include authors, population, task used, and stage(s) analyzed

	Authors	Participants	Task	Stage analyzed
Type-1 Passive Reward task				
	Guyer et al. [74]	31 adolescents (9-17) 30 adults (21-40)	passive emotion task	feedback
	Van-Leijenhorst [73]	18 adolescents (14-15) 15 adults (18-23) 17 children (10-12)	Slot Machine task	cue/anticipation
Type-2 Reward Test task				
	Van Leijenhorst [76]	14 adults (18-26) 12 children (9-12)	Cake Gambling task	selection and feedback
	Galvan et al.[75]	13 adolescents (13-17) 12 adults (23-29) 16 children (7-11)	Pirate task	whole trial
	Bjork et al. [82]	12 adolescents (12-17) 12 adults (22-28)	Monetary Incentive Delay task	anticipation and feedback
	Bjork et al. [83]	24 adolescents (12-17) 24 adults (22-42)	Monetary Incentive Delay task	anticipation and feedback
	Smith et al. [84]	35 adolescents (10-17) 35 adults (18-43)	continuous performance task	whole trial
	Geier et al. [78]	18 adolescents (13-17) 16 adults (18-30)	Antisaccade task	cue/anticipation, and cognitive/motor
	Somerville et al. [85]	19 adolescents (13-17) 25 adults (18-29) 18 children (6-12)	Go/No-go task	cue
	Cohen et al. [36]	16 adolescents (14-19) 11 adults (25-30) 18 children (8-12)	Probabilistic Learning	selection and feedback
Type-3 Reward Decision-Making task				
	Eshel et al. [86]	16 adolescents (9-17) 14 adults (20-40)	Wheel of Fortune task	selection
	Ernst et al. [87]	18 adolescents (9-17) 16 adults (20-40)	Wheel of Fortune task	selection and feedback
	Van Leijenhorst et al. [88]	15 pubertal adolescents (12-14) 15 post-pubertal adolescents (16-17) 15 adults (19-26) 13 children (8-10)	Modified Cake Gambling task	selection and feedback
	Chein et al. [89]	14 adolescents (14-18) 14 young adults (19-22) 12 adults (24-29)	Stoplight task	selection stage
	Christakou et al. [90]	19 male adolescents (11-17) 21 male adults (18-31)	Hypothetical Delay Discounting	selection stage

paradigms or *performance preparation stage* for the reward test-task, a *cognitive/motor stage* during which the required response is executed, an *anticipation stage* (participants wait to find out the outcome of their action), and a reward *feedback stage*. Rarely are the trials analyzed as *whole trials*, without separating the different stages within trials (e.g., cue-anticipation from feedback).

Passive exposure to incentive (positive/negative) stimuli: As mentioned previously, Guyer et al. [74] utilized a passive emotion task, which presented affect-laden stimuli to adolescents and adults. Although not examined from the perspective of reward processes, the presentation of negatively valenced stimuli (fearful face), a potential index of response to punishment, indicated greater amygdala activation (MNI 16, −4, −16; −20, −8, −6) in adolescents compared to adults. No significant

Table 8.3 Main results from fMRI reward studies showing between group differences, organized by type of reward task, stage analyzed, and brain areas activated (light gray: reward system, middle gray: avoidance system, and dark gray: control system). MNI and Talaraich coordinates included when available. Initials of first authors and year of study is indicated in the table and the key for these initials is provided below the table

		Cue appraisal/Selection/Performance Preparation	Cognitive/Motor performance execution	Feedback	Whole trial
Passive Reward task	Striatum			Favor: **Adol. > Adult** (VL-09); MNI 12, 9, -15	
	OFC			Unfav: **Adult > Adol.** (VL-09); MNI -27, 48,-3	
	Amygdala			Unfav: **Adol. > Adult** (G-08); MNI 16, -4, -16; -20, -8, -6	
	Insula	Favor: **Adol. > Adult** (VL-09); MNI 42, 12, -3			
	ACC				
	IFG				
	DLPFC				
Reward Test task	Striatum	Favor: **Adult > Adol.** (B-04;B-10); Tal (B-04) -9, 10, 0; 11, 12, 0; Favor: **Adol. > Adult** (Ge-09); Tal 11, 8, -7		Decision Value: **Adol.> Adult = Child** (C-10); MNI 14, 16, 4	Favor: **Adol. > Adult/Child** (G-06)
		Favor: **Adult > Adol.** (Ge-09); Tal 14, 2, -7			Favor: **Adol. > Adult** (Sm-11); Tal 29, -7, 4
		Favor (happy face): **Adol. > Adult/Child** (S-10); Tal -4, 11, -9			
	OFC		Unfav (neutral) **Adol. > Adult** (Ge-09); Tal -25, 44, -4	Unfav (Lowest prob): **Child > Adult** (VL-06); MNI 40, 46, -12	Favor: Child > **Adol.=Adult** (G-06)
	Amygdala				
	Insula				
	ACC/mPFC	Decision Value: **Adol. > Adult** (C-10); MNI 0, 50, -8			
		Unfav. **Child > Adult** (VL-06): MNI 0, 6, 20			
	IFG	Unfav (calm vs happy) **Child > Adol. > Adult** (S-10); Tal 32, 23, 3			
	DLPFC				Favor: **Adol. > Adult** (Sm-11); Tal 40, 44, 15
Reward Decision-Making task	Striatum	Social > Nonsocial: **Adol. > Adult** (Che-11); MNI 9, 12, -8		Favor: **Adol. > Adult** (E-05); MNI -16, 20, -4	
		Immediate > Delay: **Adol. > Adult** (Chr-11); Tal -7, 26, -13		Favor: **Adol. > Adult/Child** (VL-10); MNI 21, 18, 9	
	OFC	Risky: **Adult > Adol.** (Es-06); MNI -44, 14, -4			
		Social > Nonsocial: **Adol. > Adult** (Che-11); MNI -22, 47, -10			
	Amygdala			Favor: **Adult > Adol.** (E-05); MNI -26, -4, -14	
	Insula				
	ACC/mPFC	Risky: **Adult > Adol.** (Es-06): MNI ±2, 26, 30			
		Risky: **Adol. > Adult** (VL-10); MNI 12, 9, 27			
		Lower cost: **Adol. > Adult** (Chr-11); Tal -18, 46, -6			
	IFG				
	LPFC	All: **Adult > Adol.** (Che-11); MNI -31, 5, 56			

Table 8.3 continued

Favor	favorable, such as reward, appetitive stimuli	
Unfav	unfavorable, such as punishment	
Passive Reward Tasks		
G-08	Guyer et al. [74]	passive emotion task
VL-09	Van Leijenhorst et al. [73]	Slot Machine task
Reward Test Tasks		
VL-06	Van Leijenhorst et al. [76]	Cake Gambling task
G-06	Galvan et al. [75]	Pirate task
B-04	Bjork et al. [82]	Monetary Incentive Delay
B-10	Bjork et al. [83]	Monetary Incentive Delay
Sm-11	Smith et al. [84]	continuous performance task
Ge-09	Geier et al. [78]	Antisaccade task
S-10	Sommerville et al. [85]	Go/No-go task
C-10	Cohen et al. [36]	Probabilistic Learning
Reward Decision-Making Tasks		
Es-06	Eshel et al. [86]	Wheel of Fortune task
E-05	Ernst et al. [87]	Wheel of Fortune task
VL-10	Van Leijenhorst et al. [88]	Modified Cake Gambling task
Che-11	Chein et al. [89]	Stoplight task
Chr-11	Christakou et al. [90]	Hypothetical Delay Discounting task

age effect on neural activation was found in response to the presentation of appetitive stimuli (happy faces) versus baseline. The Slot Machine task [73] passively presented probabilistic rewards without behavioral correlates. This task provided a *cue/anticipation stage* and a *feedback stage*. In *anticipation* of a probabilistic reward, a linear relationship was found between age and the right anterior insula activation (MNI 42, 12, −3). Insula activation in anticipation of a probabilistic reward (vs. no reward) was significant in children and adolescents but not in adults. No age-group differences in activation were found in the striatum. The *feedback stage* showed greater activation of the striatum (MNI 12, 9, −15) to reward receipt in adolescents than adults or children, and greater activation of the orbitofrontal cortex (OFC; MNI −27, 48,-3) in adults than adolescents or children to reward omission.

Taken together (see Table 8.3), these studies suggest that when probabilistic rewards are presented passively, the insula, within the *emotion module*, seems to be more responsive in adolescents than in adults during reward cue appraisal. The *reward module* (striatum and OFC), on the other hand, shows a dissociation in the region and age effect in response to stimulus valence: striatal activation is greater in adolescents than in adults in response to appetitive stimuli, but OFC activation is greater in adults than in adolescents in response to an aversive stimulus. Together with this group effect on OFC in the passive exposure to a negative stimulus, the amygdala is more activated in adolescents than in adults. One possible interpretation is that OFC/vPFC serves to modulate the amygdala during aversive stimulus in adults, whereas OFC is less recruited in adolescents, whose amygdala, in turn, is not inhibited and more active than the adult amygdala.

Reward Test-Tasks Most reward-related paradigms that have been used in developmental studies have been reward test-tasks. In the Cake Gambling task [76], subjects are asked to select the “best”, or correct, probability for winning points. The effects of age on neural activation in response to the Cake Gambling task were two-fold: during the *selection stage*, the medial PFC/anterior cingulate cortex (ACC; MNI 0, 6, 20) was activated more in children than in adults to lower probability outcomes (i.e., incorrect responses); and during the *feedback stage*, the OFC (MNI 40, 46, −12) was more sensitive to negative feedback in children than in adults. This OFC finding is opposite to the OFC finding described above in passive reward tasks. Differences between studies include the type of task, the stage of reward process being examined (cue appraisal vs. feedback), and the age of the young sample (children vs. adolescents).

In the Pirate task, a delayed response task with two possible response options [75], subjects were asked to select the side of the screen where the Pirate was just presented, and they were rewarded for selecting the correct response. During the *feedback stage* of this task, adolescents showed greater nucleus accumbens activation compared with children and adults and lower OFC activation (peak coordinates not available) in response to correct trials than children, but no significant difference from adults across the *whole trial*.

Two studies [82, 83] used the Monetary Incentive Delay (MID) task, which is a reaction-time test paired with different gain and loss levels. In the first study, during the *anticipation stage* (in preparation for the test), the main finding was reduced ventral striatal activation (Tal −9, 10, 0; 11, 12, 0) in adolescents compared to adults, when potential gains were at stake. During the *feedback stage*, both age groups similarly activated the PFC, nucleus accumbens, putamen, amygdala and hippocampus in response to gains. In response to non-losses vs. losses, both groups activated putamen (deactivation to losses), and only adults activated medial PFC (Tal 1, 53, −6). Overall, the group differences were significant only for the ventral striatum during the *anticipation stage*.

The second study by Bjork and colleagues [83] was a replication with a larger sample and some modifications of the timing parameters to better dissociate the *anticipation stage* from the *feedback stage*. The overall pattern of activation was replicated. The only significant group difference was lower activation of the nucleus accumbens response to both cue-gain anticipation and to cue-loss anticipation in adolescents than in adults.

Researchers have also examined incentive effects on various cognitive tests, including tests of inhibition [78, 85], sustained-attention [84] and learning [91]. For instance, Smith et al. [84] employed a continuous performance task (CPT) with 3 types of trials: non-targets, rewarded targets and non-rewarded targets. Behaviorally, adolescents evidenced significantly slower responding to nonrewarded targets relative to adults. Adolescents also responded significantly faster to rewarded targets relative to nonrewarded targets, while no such difference was found among adults, suggesting that the effects of incentives on sustained attention is stronger in adolescents than adults. Regarding neural functioning, the comparison of neural activation in response to rewarded vs. non-rewarded targets revealed positive linear relationships between age and reward-induced activation in regions implicated in sustained attention (DLPFC, Tal 40, 44, 15; ventromedial OFC, Tal −29, 26,

−2), but negative linear relationships between age and reward-induced activation in regions coding for visuospatial attention (R putamen, Tal 29, −7, 4; posterior cingulate cortex, Tal −14, −37, 20; inferior temporal gyrus, Tal −14, −37, 20). This finding suggests that age modulates differently the effect of reward on sustained attention and visuospatial attention. From the perspective of the dual attention model, sustained attention recruits predominantly the intention-guided attentional network, whereas visuospatial attention refers more specifically to the stimulus-guided attentional network [92]. This age-related dissociation of the effect of reward on these two networks may reflect a facilitation of stimulus-driven processes in youth, but of intention-driven processes in adults [93].

Geier et al. [78] examined the effects of reward on antisaccades, during the *appraisal-cue stage* and *saccade preparation* and *saccade execution (*cognitive/motor stage). An antisaccade is an eye movement to the opposite direction of a suddenly appearing target. This action requires the inhibition of a prepotent response toward the target, and the execution of an endogenously guided response in the opposite direction of the suddenly appearing stimulus. A time-course analysis over 18 seconds post-trial onset was used to assess neural responses during the three stages of *cue*, saccade *anticipation*, and saccade *execution*. During the incentive *cue*, the ventral striatum was recruited more strongly in adults than in adolescents (Tal 14, 2, −7), during saccade *anticipation*, it was recruited more strongly in adolescents than in adults (Tal 11, 8, −7), and saccade execution (*cognitive/motor stage*) showed no group differences. The OFC showed a group difference only during saccade *execution*, with a stronger response to neutral cues in adolescents than adults (Tal −25, 44, −4). Regions recruited in antisaccade tasks tended to be more activated in adolescents than adults in response to neutral trials, but showed no group differences to reward trials. In sum, this study showed a more responsive reward system in adolescents during anticipation for action (here mostly inhibition) during reward trials. The other stages showed group differences during the neutral trials, which disappeared when incentives were present, suggesting "normalization" by incentives of neural response in adolescents to match that in adults.

Somerville et al. [85] examined the effects of positive "emotion" rather than positive reinforcement (i.e., reward). We nevertheless include this study here as it was interpreted along the framework of reward processes. A go/no-go paradigm was used, with facial emotions as stimuli. Happy or calm expressions were alternatively used as go or no-go stimuli in separate blocks. Of interest, behaviorally, all 3 age groups (children, adolescents and adults) showed faster reaction time to happy than to calm faces. In addition, adolescents had more false alarms to happy faces than did adults or children. At the neural level, adolescents engaged the ventral striatum (Tal −4, 11, −9) more strongly than either adults or children, particularly to happy faces. Activation of the key inhibitory brain area, the inferior frontal gyrus (IFG; Tal 32, 23, 3), showed linear decreased activation to the contrast "no-go vs. go" across all faces with age, and greater activation to the contrast "calm vs. happy" across groups. This work suggests that the ventral striatum response to positive, social appetitive stimuli is greater in adolescents than either adults or children, which parallels the facilitation of behavioral approach (false alarms) to happy faces in adolescents relative to adults or children.

The last task of probabilistic learning by Cohen et al. [91] probed learning through repetition. With increasing experience, subjects learned which of two options was most frequently correct (and rewarded). Since this task did not involve risk-taking, but examined the effect of reward on a cognitive task (learning), we decided to include it in the test-task section rather than in the category of decision-making tasks, although we recognize that this choice is debatable. In this task, participants were shown pairs of abstract stimuli, and were asked to classify them into two categories (i.e., Eastern or Northern). Feedback as to whether their response was correct was given after each trial. Two types of stimuli were presented, two were predictable (associated at a rate of 83 % with a given category), two were random (associated at a rate of 50 % with each category). There were also 2 levels of reward, high ($0.25) and low ($0.05). During the *feedback stage*, activation in response to positive prediction errors (unexpected gain) peaked in the striatum (MNI 14, 16, 4) and angular gyrus (MNI 60, −44, 28) in adolescents relative to adults or children. In addition, decision value, i.e., the value assigned to each potential choice [94], during the *selection stage* was associated with a linear decrease of medial PFC activation with age (MNI 0, 50, −8). Of note, behaviorally, the estimated learning rate did not differ among age-groups.

Taken together (see Table 8.3), these reward-test studies showed that during *cue appraisal*, striatal activation was lower in adolescents than in adults on reward trials in two different tasks [78, 82, 83]. However, during *preparation for test performance* (e.g., inhibition of an ocular response as in [78], or of a manual response as in [85]), striatal activation during reward trials was greater in adolescents than in adults. This was also the case when the test required learning values [91]. During actual *test performance*, prefrontal cortical regions (IFG and DLPFC) were more active in youths than in adults on non-rewarded trials [76, 85]. Similarly, *test performance* (antisaccade here [78]) during a neutral trial (not rewarded) was accompanied with greater OFC activation in adolescents than in adults. Smith and colleagues [84], on the other hand, reported greater DLPFC activation with age on rewarded versus nonrewarded target trials. Analysis of the trials as a whole revealed greater striatal activation in positive trials in adolescents than in adults, but lower OFC activation in children than in adults, with the adolescents not differing from the adults [75].

Overall, generally consistent data seem to emerge: during the cue-appraisal stage of the type of trial, adolescents activate the striatum less than adults, but during the performance preparation stage (i.e., preparation for action), adolescents activate striatum more than adults, particularly in reward-trials. The distinction between these two early stages of reward processes, which has not been explicitly noted in the past, appears to be quite important based on the differential relative reliance on striatal function by adolescents and adults. More work is needed to validate this observation. Finally, during positive *feedback stage*, stronger striatal activation [91], and, during negative feedback, weaker OFC activation [76] emerged in adolescents compared to adults. Of interest, these reward test-tasks do not seem to modulate amygdala or insula differentially as a function of age.

Decision-Making Tasks Four decision-making paradigms have been examined in different age-groups, the Wheel of Fortune (WOF) task [86, 87], the Stoplight task [89], a temporal discounting task (TDT) [90], and the modified Cake Gambling task [88].

In the WOF task, participants were asked to select one of two options, which varied by magnitude and probability of monetary gains. Three stages were modeled, including *selection* (including cue appraisal), *anticipation* and *feedback*. Only *selection* and *feedback* were analyzed. During the *selection stage*, regions of the OFC/ventrolateral PFC (MNI −44, 14, −4) and ACC (MNI ±2, 26, 30) were significantly more engaged in adults than in adolescents, when selecting the most risky (most uncertain but potentially most lucrative) option [86]. Other regions, such as ventral striatum, amygdala and DLPFC, were also activated by this contrast, but in a similar way for both age groups [87]. In contrast, the *feedback stage* was accompanied by significant age-differences in activation of the ventral striatum and amygdala. The amygdala (MNI −26, −4, −14) tended to be more activated in adults than in adolescents in response to gain vs. no-gain outcome (greater deactivation to no-gain). In contrast, the nucleus accumbens (MNI −16, 20, −4), within the ventral striatum, was more activated in adolescents compared to adults (higher activation to gain). Taken together, these findings support a hypersensitivity of the reward neural system in adolescence and a reduction of conflict-related ACC engagement in adolescents relative to adults, in a task of reward-related decision-making.

In the modified Cake Gambling task, participants were asked to select among two options, each with a fixed probability of 33 % and 66 % [88]. Trials differed on the potential gain associated with the 33 % (risky) option (2, 4, 6 or 8 €), while the 66 % (safe) option always provided a potential 1 € gain. The selection (also including cue-appraisal) and feedback stages were analyzed separately. During the *selection stage* of risky (33 % gain probability) high incentive vs. low incentive options, a linear decrease with age was found in dorsal ACC (MNI 12, 9, 27) and central opercular postcentral gyrus (MNI 51, −6, 21). In contrast to the previous WOF study, adolescents showed stronger activation compared to adults or children in the medial ventral PFC/subcallosal cortex (MNI −9, 27, −12). During the *feedback stage* in the gain condition, no linear associations of brain activation with age were found, but a region of significantly greater activation in adolescents relative to adults or children was detected in the right caudate nucleus (MNI 21, 18, 9). The authors concluded that the reduction of reward-related activation of the ACC with age indexed maturation of cognitive control regions, whereas the previous work [86] argued that the more mature PFC was more readily activated in adults than in adolescents. Only with additional studies will these controversies be understood. In addition, in line with previous work, the peak activation in adolescence of medial ventral PFC and striatum was attributed to the unique sensitivity of adolescents to positive incentives.

The Stoplight task is a simulated driving game that was performed by adolescents, and young and older adults [95]. Participants were asked to decide whether to risk a collision by driving through an orange light and saving time, or to stop and be safe. Completion of the task in a timely fashion was rewarded by a monetary incentive. In addition, the task was done in two contextual conditions, social when two age-matched peers supported the participant, and non-social when playing the task alone. Only the adolescents took more risks during the social than the non-social context. Risky and safe selections were collapsed together, and were compared to an implicit baseline. Adults showed greater activation than adolescents during the *selection stage* in the lateral PFC (MNI −31, 5, 56), inferior parietal (MNI −52, −37, 41), and fusiform gyrus (−52, −55, −19). No regions were more activated in adolescents than in adults in the selection vs. baseline contrast. However, in the comparison of social and nonsocial conditions, adolescents activated ventral striatum (MNI 9, 12, −8) and OFC (MNI −22, 47, −10) more strongly than adults.

A hypothetical discount task [90] assessed the subjective cost associated with a delay in reward attainment as reflected in type of selection. Performance data showed that this cost was higher in adolescents than in adults. The neuroimaging data, collected during the *selection stage*, found that this cost was associated with higher activation of ventral striatum (Tal −7, 26, −13), putamen/thalamus (Tal −22, −15, 4), and superior parietal lobule (Tal 25, −63, 53) in adolescents when selecting immediate reward, and in temporal regions during delayed choices. The only region showing increasing activation with age, and with decreased cost (i.e., larger delays), was the ventromedial PFC (Tal −18, 46, −6). In addition, connectivity analyses showed strengthening of functional links between ventromedial PFC and ventral striatum with age during selection of immediate options. The authors concluded that ventromedial PFC was progressively more able to incorporate information about the delay-dependent value of future rewards.

Collectively, these studies suggest that during a reward decision-making task (see Table 8.3), the early stage of cue-appraisal/selection is associated with greater striatal activation in adolescents than adults. However, discrepant findings in OFC and mPFC (e.g., greater activation in adults than adolescents in risky vs. non-risky decisions, and greater activation in adolescents than in adults in more social and less costly decisions) did emerge. More work is warranted to resolve these differences. During the feedback stage, favorable outcomes were associated with greater striatal activation in adolescents than adults, but greater activation in the amygdala in adults than adolescents. Thus, studies of decision-making inform the functional contributions of the three neural modules of the triadic model, and reflect the important notion that the equilibrium among these systems depend on the types of cognitive demands and reward conditions associated with the decision-making task.

Overall, these findings across all the reward-related tasks suggest mixed results that do not segregate by type of task. The scarcity and heterogeneity of studies make it difficult to clearly map the role of developmental changes through adolescence onto specific mechanisms, particularly in ways that could clearly predict the dynamic changes within the triadic model.

For completeness, a number of reward tasks have been assessed only in pediatric samples (e.g. [19, 37, 81, 96–100]). These studies suggest that the neural pathways involved in reward-related processes in youth engage similar circuits as those in adults. However, only a direct comparison of different age-groups can address age effects on the neural substrates of behavior. Of course, longitudinal studies represent the state of the art approach to examine age effects. Unfortunately, these studies are costly and difficult to conduct.

Conclusion

We showed that the modulation of the balance across the triadic model is not straightforward to interpret for a number of reasons. (1) The use of different tasks that probe different component processes and brain functions prevent a step-by-step logical testing of reward function. However, these studies now provide a large basis for starting such systematic analysis. (2) In addition, each module of the triadic system is not systematically assessed separately as well as in relation with one another across extant studies. To test the triadic model, study designs should permit one to orthogonalize the experimental manipulations of the approach, avoidance and control systems. (3) A third glowing gap is the quasi-absence of functional connectivity studies during reward-related processes. The field is still very young. Studies of resting state connectivity across development are starting to emerge [101, 102], and provide important starting points for investigations of functionally constrained studies.

Although mainly described in the context of development, the triadic model is an ideal tool to guide research on the neurobiology of vulnerability to substance abuse, particularly in the adolescent. The most parsimonious account of the propensity for substance abuse in adolescents rests on the typical changes in behavior and the associated neural alterations that occur during this period. Increased impulsivity and risk-taking have been linked to changes in cognitive control, reward function, and sensitivity to aversive stimuli [79]. Critically, these changes vary as a function of context. For example, they are exacerbated in a social relative to a non-social context [89, 103]. The triadic model provides hypothetical activity patterns of the specific functional networks (i.e., reward, avoidance, control) that underlie the characteristics of motivated behavior across adolescence, and which can lead to substance abuse. These putative functional brain maps can be tested in a priori studies, through a systematic and one-step at a time manipulation of the cognitive and contextual factors linked to substance use propensity and behavior. We hope that this review will inspire and guide such studies.

References

1. Dahl RE. Adolescent brain development: a period of vulnerabilities and opportunities. Keynote address. Ann N Y Acad Sci. 2004;1021:1–22.
2. Ernst M, Hardin M. Neurodevelopment underlying adolescent behavior a neurobiological model. In: Zelazo PD, Chandler M, & Crone E, editors. Developmental social cognitive neuroscience. New York: Lawrence Erlbaum Associates; 2009. pp. 165–89.
3. Ernst M, Paulus MP. Neurobiology of decision making: a selective review from a neurocognitive and clinical perspective. Biol Psychiatry. 2005;58(8):597–604.
4. DiClemente RJ, Lodico M, Grinstead OA, Harper G, Rickman RL, Evans PE, et al. African-American adolescents residing in high-risk urban environments do use condoms: correlates and predictors of condom use among adolescents in public housing developments. Pediatrics. 1996;98(2 Pt 1):269–78.
5. Gullo MJ, Dawe S. Impulsivity and adolescent substance use: rashly dismissed as "all-bad"? Neurosci Biobehav Rev. 2008;32(8):1507–18.
6. Windle M, Spear LP, Fuligni AJ, Angold A, Brown JD, Pine D, et al. Transitions into underage and problem drinking: developmental processes and mechanisms between 10 and 15 years of age. Pediatrics. 2008;121(Suppl 4):273–89.
7. Hingson R, Heeren T, Winter M, Wechsler H. Magnitude of alcohol-related mortality and morbidity among U.S. college students ages 18–24: changes from 1998 to 2001. Annu Rev Public Health. 2005;26:259–79.
8. Spear LP. The adolescent brain and age-related behavioral manifestations. Neurosci Biobehav Rev. 2000;24(4):417–63.
9. Costello EJ, Pine DS, Hammen C, March JS, Plotsky PM, Weissman MM, et al. Development and natural history of mood disorders. Biol Psychiatry. 2002;52(6):529–42.
10. Kessler RC, Berglund P, Demler O, Jin R, Merikangas KR, Walters EE. Lifetime prevalence and age-of-onset distributions of DSM-IV disorders in the national comorbidity survey replication. Arch Gen Psychiatry. 2005;62(6):593–602.
11. Bogacz R, McClure S, Li J, Cohen J, Montague P. Short-term memory traces for action bias in human reinforcement learning. Brain Res. 2007;1153:111–21.
12. Glimcher PW, Kable J, Kenway L. Neuroeconomic studies of impulsivity: now or just as soon as possible? Am Econ Rev. 2007;97(2):142–7.
13. Hariri AR, Brown SM, Williamson DE, Flory JD, de Wit H, Manuck SB. Preference for immediate over delayed rewards is associated with magnitude of ventral striatal activity. J Neurosci. 2006;26(51):13213–7.
14. Roesch MR, Calu DJ, Burke KA, Schoenbaum G. Should I stay or should I go? Transformation of time-discounted rewards in orbitofrontal cortex and associated brain circuits. Ann N Y Acad Sci. 2007;1104:21–34.
15. Andrews-Hanna JR, Mackiewicz Seghete KL, Clause ED, Burgess GC, Ruzic L, Banich MT. Cognitive control in adolescence: neural underpinnings and relation to seld-report behaviors. PLoS ONE. 2011;6(6):e21598.
16. Boyer TW. The development of risk-taking: a multi-perspective review. Dev Rev. 2006;26(3):291–345.
17. Steinberg L. Risk taking in adolescence: what changes, and why? Ann N Y Acad Sci. 2004;1021:51–8.
18. Leigh BC. Peril, chance, adventure: concepts of risk, alcohol use and risky behavior in young adults. Addiction. 1999;94(3):371–83.
19. Crone EA, van der Molen MW. Developmental changes in real life decision making: performance on a gambling task previously shown to depend on the ventromedial prefrontal cortex. Dev Neuropsychol. 2004;25(3):251–79.
20. Ernst M, Pine DS, Hardin M. Triadic model of the neurobiology of motivated behavior in adolescence. Psychol Med. 2006;36(3):299–312.

21. Lejuez CW. Evaluation of the balloon analogue risk task (BART) as a predictor of adolescent real-world risk-taking behaviours. J Adolesc. 2003;26(4):475–9.
22. Lejuez CW, Read JP, Kahler CW, Richards JB, Ramsey SE, Stuart GL, et al. Evaluation of a behavioral measure of risk taking: the balloon analogue risk task (BART). J Exp Psychol Appl. 2002;8(2):75–84.
23. Lejuez CW, Aklin WM, Daughters S, Zvolensky MJ, Kahler CW, Gwadz M. Reliability and validity of the youth version of the balloon analogue risk task (BART-Y) in the assessment of risk-taking behavior among inner-city adolescents. Clin Child Adolesc Psychol. 2007;36(1):106–11.
24. Overman WH, Frassrand K, Ansel S, Trawalter S, Bies B, Redmond A. Performance on the IOWA card task by adolescents and adults. Neuropsychologia. 2004;42(13):1838–51.
25. Rao H, Korczykowski M, Pluta J, Hoang A, Detre J. Neural correlates of voluntary and involuntary risk taking in the human brain: an fMRI study of the Balloon Analog Risk Task (BART). Neuroimage. 2008;42(2):902–10.
26. Steinberg L, Albert D, Cauffman E, Banich M, Graham S, Woolard J. Age differences in affective decision making as indexed by performance on the Iowa Gambling Task. Dev Psychol. 2010;46(1):193–207.
27. Aklin WM, Lejuez CW, Zvolensky MJ, Kahler CW, Gwadz M. Evaluation of behavioral measures of risk taking propensity with inner city adolescents. Behav Res Ther. 2005;43(2):215–28.
28. Lejuez CW, Aklin WM, Zvolensky MJ, Pedulla CM. Evaluation of the balloon analogue risk task (BART) as a predictor of adolescent real-world risk-taking behaviours. J Adolesc. 2003;26(4):475–9.
29. Deakin J, Aitken M, Robbins T, Sahakian BJ. Risk taking during decision-making in normal volunteers changes with age. J Int Neuropsychol Soc. 2004;10(4):590–8.
30. Figner B, Mackinlay RJ, Wilkening F, Weber E. Supplemental material for affective and deliberative processes in risky choice: age differences in risk taking in the Columbia card task. J Exp Psychol Learn Mem Cognit. 2009;35(3):709–30.
31. Hooper CJ, Luciana M, Conklin HM, Yarger RS. Adolescents' performance on the Iowa Gambling Task: implications for the development of decision making and ventromedial prefrontal cortex. Dev Psychol. 2004;40(6):1148–58.
32. Steinberg L. A dual systems model of adolescent risk-taking. Dev Psychobiol. 2010. doi:10.1002/dev.20445.
33. Silk JS, Forbes EE, Whalen DJ, Jakubcak JL, Thompson WK, Ryan ND, et al. Daily emotional dynamics in depressed youth: a cell phone ecological momentary assessment study. J Exp Child Psychol. 2010. doi:10.1016/j.jecp.2010.10.007.
34. Brown SA, Tapert SF. Adolescence and the trajectory of alcohol use: basic to clinical studies. Ann N Y Acad Sci. 2004;1021:234–44.
35. Gardner M, Steinberg L. Peer influence on risk taking, risk preference, and risky decision making in adolescence and adulthood: an experimental study. Dev Psychol. 2005;41(4):625–35.
36. Cohen GL, Prinstein MJ. Peer contagion of aggression and health risk behavior among adolescent males: an experimental investigation of effects on public conduct and private attitudes. Child Dev. 2006;77(4):967–83.
37. Casey BJ, Jones RM, Hare TA. The adolescent brain. Annal New York Acad Sci. 2007;1124(1):111–26.
38. Giedd JN, Blumenthal J, Jeffries NO, Castellanos FX, Liu H, Zijdenbos A, et al. Brain development during childhood and adolescence: a longitudinal MRI study. Nat Neurosci. 1999;2(10):861–3.
39. Rubia K, Smith AB, Woolley J, Nosarti C, Heyman I, Taylor E, et al. Progressive increase of frontostriatal brain activation from childhood to adulthood during event-related tasks of cognitive control. Hum Brain Mapp. 2006;27(12):973–93.
40. Toga AW, Thompson PM, Sowell ER. Mapping brain maturation. Trends Neurosci. 2006;29(3):148–59.

41. Giedd JN. Structural magnetic resonance imaging of the adolescent brain. Ann N Y Acad Sci. 2004;1021:77–85.
42. Gogtay N, Giedd JN, Lusk L, Hayashi KM, Greenstein D, Vaituzis AC, et al. Dynamic mapping of human cortical development during childhood through early adulthood. Proc Natl Acad Sci U S A. 2004;101(21):8174–9.
43. Paus T, Zijdenbos A, Worsley K, Collins DL, Blumenthal J, Giedd J, et al. Structural maturation of neural pathways in children and adolescents: in vivo study. Science. 1999;283(5409):1908–11.
44. Paus T, Keshavan M, Giedd J. Why do many psychiatric disorders emerge during adolescence? Nat Rev Neurosci. 2008;9(12):947–57.
45. Ernst M, Fudge JL. A developmental neurobiological model of motivated behavior: Anatomy, connectivity and ontogeny of the triadic nodes. Neurosci Biobehav Rev. 2009;33(3):367–82.
46. Di Chiara G, Bassareo V. Reward system and addiction: what dopamine does and doesn't do. Curr Opin Pharmacol. 2007;7(1):69–76.
47. Kringelbach ML. The human orbitofrontal cortex: linking reward to hedonic experience. Nat Rev Neurosci. 2005;6(9):691–702.
48. Wise RA. Dopamine, learning and motivation. Nat Rev Neurosci. 2004;5(6):483–94.
49. Davis M. Neural systems involved in fear and anxiety measured with fear-potentiated startle. Am Psychol. 2006;61(8):741–56.
50. LeDoux JE. Emotion circuits in the brain. Annu Rev Neurosci. 2000;23:155–84.
51. Pine DS. Research review: a neuroscience framework for pediatric anxiety disorders. J Child Psychol Psychiatry. 2007;48(7):631–48.
52. Rauch SL, Shin LM, Wright CI. Neuroimaging studies of amygdala function in anxiety disorders. Ann N Y Acad Sci. 2003;985:389–410.
53. Amodio DM, Frith CD. Meeting of minds: the medial frontal cortex and social cognition. Nat Rev Neurosci. 2006;7(4):268–77.
54. Aron AR, Robbins TW, Poldrack RA. Inhibition and the right inferior frontal cortex. Trends Cogn Sci. 2004;8(4):170–7.
55. Bush G, Luu P, Posner MI. Cognitive and emotional influences in anterior cingulate cortex. Trends Cogn Sci. 2000;4(6):215–22.
56. Carter CS, van Veen V. Anterior cingulate cortex and conflict detection: an update of theory and data. Cogn Affect Behav Neurosci. 2007;7(4):367–79.
57. Ridderinkhof KR, van den Wildenberg WP, Segalowitz SJ, Carter CS. Neurocognitive mechanisms of cognitive control: the role of prefrontal cortex in action selection, response inhibition, performance monitoring, and reward-based learning. Brain Cogn. 2004;56(2):129–40.
58. Blakemore SJ. The social brain in adolescence. Nat Rev Neurosci. 2008;9(4):267–77.
59. Nelson EE, Leibenluft E, McClure EB, Pine DS. The social re-orientation of adolescence: a neuroscience perspective on the process and its relation to psychopathology. Psychol Med. 2005;35(2):163–74.
60. Jensen J, McIntosh AR, Crawley AP, Mikulis DJ, Remington G, Kapur S. Direct activation of the ventral striatum in anticipation of aversive stimuli. Neuron. 2003;40(6):1251–7.
61. Ernst M, Spear LP. Reward systems. In: Haan M de, Gunnar MR, editors. Handbook of developmental social neuroscience. vol. xvii. New York: Guilford Press; 2009. p. 558.
62. Hardin MG, Pine DS, Ernst M. The influence of context valence in the neural coding of monetary outcomes. Neuroimage. 2009b;48(1):249–57.
63. Arnett JJ. Adolescent storm and stress, reconsidered. Am Psychol. 1999;54(5):317–26.
64. Larson RW, Moneta G, Richards MH, Wilson S. Continuity, stability, and change in daily emotional experience across adolescence. Child Dev. 2002;73(4):1151–65.
65. Silk JS, Steinberg L, Morris AS. Adolescents' emotion regulation in daily life: links to depressive symptoms and problem behavior. Child Dev. 2003;74(6):1869–80.
66. Weinstein SM, Mermelstein RJ, Hankin BL, Hedeker D, Flay BR. Longitudinal patterns of daily affect and global mood during adolescence. J Res Adolesc. 2007;17(3):587–600.

67. Chikazoe J, Konishi S, Asari T, Jimura K, Miyashita Y. Activation of right inferior frontal gyrus during response inhibition across response modalities. J Cogn Neurosci. 2007;19(1):69–80.
68. Liddle PF, Kiehl KA, Smith AM. Event-related fMRI study of response inhibition. Hum Brain Mapp. 2001;12(2):100–9.
69. Rubia K, Hyde Z, Halari R, Giampietro V, Smith A. Effects of age and sex on developmental neural networks of visual-spatial attention allocation. Neuroimage. 2010;51(2):817–27.
70. Marsh R, Zhu H, Schultz RT, Quackenbush G, Royal J, Skudlarski P, et al. A developmental fMRI study of self-regulatory control. Hum Brain Mapp. 2006;27(11):848–63.
71. Rubia K, Smith AB, Taylor E, Brammer M. Linear age-correlated functional development of right inferior fronto-striato-cerebellar networks during response inhibition and anterior cingulate during error-related processes. Hum Brain Mapp. 2007b;28(11):1163–77.
72. Leshner AI. Addiction is a brain disease, and it matters. Science. 1997;278(5335):45–7.
73. Van Leijenhorst L, Zanolie K, Van Meel CS, Westenberg PM, Rombouts SARB, Crone EA. What motivates the adolescent? Brain regions mediating reward sensitivity across adolescence. Cereb Cortex. 2009;20(1):61–9.
74. Guyer AE, Monk CS, McClure-Tone EB, Nelson EE, Roberson-Nay R, Adler AD, et al. A developmental examination of amygdala response to facial expressions. J Cogn Neurosci. 2008;20(9):1565–82.
75. Galvan A. Earlier Development of the accumbens relative to orbitofrontal cortex might underlie risk-taking behavior in adolescents. J Neurosci. 2006;26(25):6885–92.
76. Van Leijenhorst L, Crone E, Bunge S. Neural correlates of developmental differences in risk estimation and feedback processing. Neuropsychologia. 2006;44(11):2158–70.
77. Knutson B, Fong GW, Adams CM, Varner JL, Hommer D. Dissociation of reward anticipation and outcome with event-related fMRI. Neuroreport. 2001;12(17):3683–7.
78. Geier CF, Terwilliger R, Teslovich T, Velanova K, Luna B. Immaturities in reward processing and its influence on inhibitory control in. Adolesc Cereb Cortex. 2009;20(7):1613–29.
79. Hardin MG, Mandell D, Mueller SC, Dahl RE, Pine DS, Ernst M. Inhibitory control in anxious and healthy adolescents is modulated by incentive and incidental affective stimuli. J Child Psychol Psychiatry. 2009a;50(12):1550–8.
80. Jazbec S, McClure E, Hardin M, Pine DS, Ernst M. Cognitive control under contingencies in anxious and depressed adolescents: an antisaccade task. Biol Psychiatry. 2005;58(8):632–9.
81. Bar-Haim Y, Fox NA, Benson B, Guyer AE, Williams A, Nelson EE, et al. Neural correlates of reward processing in adolescents with a history of inhibited temperament. Psychol Sci. 2009;20(8):1009–18.
82. Bjork JM, Knuton B, Fong GW, Caggiano DM, Bennett SM, et al. Incentive-elicited brain activation in adolescents: similarities and differences from young adults. J Neurosci. 2004;24(8):1793–802.
83. Bjork JM, Smith AR, Chen G, Hommer DW. Adolescents, adults and rewards: comparing motivational neurocircuitry recruitment using fMRI. PLoS ONE. 2010;5(7):e11440.
84. Smith AB, Halari R, Giampetro V, Brammer M, Rubia K. Developmental effects of reward on sustained attention networks. NeuroImage. 2011. doi:10.1016/j.neuroimage.2011.01.072.
85. Somerville LH, Casey BJ. Developmental neurobiology of cognitive control and motivational systems. Curr Opin Neurobiol. 2010;20(2):236–41.
86. Eshel N, Nelson EE, Blair RJ, Pine DS, Ernst M. Neural substrates of choice selection in adults and adolescents: development of the ventrolateral prefrontal and anterior cingulate cortices. Neuropsychologia. 2007;45(6):1270–9.
87. Ernst M, Nelson E, Jazbec S, McClure E, Monk C, Leibenluft E, et al. Amygdala and nucleus accumbens in responses to receipt and omission of gains in adults and adolescents. Neuroimage. 2005;25(4):1279–91.
88. Van Leijenhorst L, Moor BG, Op de Macks ZA, Rombouts SARB, Westenberg PM, Crone EA. Adolescent risky decision-making: neurocognitive development of reward and control regions. Neuroimage. 2010;51(1):345–55.

89. Chein J, Albert D, O'Brien L, Uckert K, Steinberg L Peers increase adolescent risk taking by enhancing activity in the brain's reward circuitry. Dev Sci. 2011;14(2):F1–10.
90. Christakou A, Brammer M, Rubia K. Maturation of limbic corticostriatal activation and connectivity associated with developmental changes in temporal discounting. Neuroimage. 2011;54(2):1344–54.
91. Cohen JR, Asarnow RF, Sabb FW, Bilder RM, Bookheimer SY, Knowlton BJ, et al. A unique adolescent response to reward prediction errors. Nat Neurosci. 2010;13(6):669–71.
92. Corbetta M, Shulman GL. Control of goal-directed and stimulus-driven attention in the brain. Nat Rev Neurosci. 2002;3:201–15.
93. Ernst M, Daniele T, Frantz KJ. New perspectives on adolescent motivated behavior: attention and conditioning. Dev Cognit Neurosci. in press. doi:10.1016/j.dcn.2011.07.013.
94. Kahneman D, Tversky A. Prospect theory: an analysis of decision under risk. Exonometrica. 1979;47:263–91.
95. Chein J, Albert D, O'Brien L, Uckert K, Steinberg L. Peers increase adolescent risk taking by enhancing activity in the brain's reward circuitry. Dev Sci. 2010;14(2):F1–10.
96. Crowley MJ, Wu J, Crutcher C, Bailey CA, Lejuez CW, Mayes LC. Risk-taking and the feedback negativity response to loss among at-risk adolescents. Dev Neurosci. 2009;31(1–2):137–48.
97. Forbes EE, Ryan ND, Phillips ML, Manuck SB, Worthman CM, Mayles DL, et al. Healthy adolescents' neural response to reward: associations with puberty, positive affect, and depressive symptoms. J Am Acad Child Adolesc Psychiatry. 2010;49(2):162–72.
98. Guyer AE, Nelson EE, Perez-Edgar K, Hardin M, Roberson-Nay R, Monk CS, et al. Striatal functional alteration in adolescents characterized by early childhood behavioral inhibition. J Neurosci. 2006;26(24):6399–405.
99. May JC, Delgado MR, Dahl RE, Stenger VA, Ryan ND, Fiez JA, et al. Event-related functional magnetic resonance imaging of reward-related brain circuitry in children and adolescents. Biol Psychiatry. 2004;55(4):359–66.
100. Rubia K, Smith A, Taylor E. Performance of children with attention deficit hyperactivity disorder (ADHD) on a test battery of impulsiveness. Child Neuropsychol. 2007a;13(3):276–304.
101. Fair DA, Cohen AL, Power JD, Dosenbach NU, Church JA, Miezin FM, et al. Functional brain networks develop from a "local to distributed" organization. PLoS Comput Biol. 2009;5(5):e1000381.
102. Stevens MC, Pearlson GD, Calhoun VD. Changes in the interaction of resting-state neural networks from adolescence to adulthood. Hum Brain Mapp. 2009;30(8):2356–66.
103. Steinberg L. A social neuroscience perspective on adolescent risk-taking. Dev Rev. 2008;28(1):78–106.

Rista C. Plate Department of Psychology, University of Wisconsin-Madison. Clinical Psychology

Jessica M. Richards Johns Hopkins University School of Medicine, Department of Psychiatry and Human Behavior. Clinical Psychology PhD.

Monique Ernst MD, PhD National Institute of Mental Health, Intramural Research Program, Section of Development and Affective Neuroscience. MD, Neurophysiology PhD.

Chapter 9
Sleep and Drug Abuse during Adolescence

Gabriel Natan Pires, Monica Levy Andersen and Sergio Tufik

Introduction

The relationship between sleep and drug abuse in adolescents is a complex topic, mostly because both phenomena are multifaceted by nature. Indeed, the present relationship do not merely restrict the association of sleep and drug abuse to a specific age group; both substance abuse and sleep patterns during adolescence are specific to this age group and differ from those observed in other life periods. Thus, restricting an approach to adolescents involves addressing a highly specific topic whose conclusions may not be extrapolated to other age groups.

Several authors argue that the relationship between sleep and drug abuse during adolescence is bidirectional [1, 2]. This statement is true: Substance use during adolescence predisposes individuals to sleep-related complaints, and adolescents who have sleep disorders are more likely to use psychoactive drugs. However, the mechanisms involved on each direction of this relationship are different. The effects of drug use on sleep patterns are mostly due to neurobiological reasons, being related to the characteristics of the drug and its pattern of use. Thus, conclusions regarding this direction of the relationship may not be specific to adolescence; being also valid for other age groups. In contrast, the finding that sleep-related complaints and sleep disorders predispose individuals to drug use has no solid neurobiological basis; rather, it is best explained by social and behavioral factors. In this case, the relationship is specific to adolescence because of their particular behavioral phenotype and social context; thus, these findings cannot be generalized to other age groups (Fig. 9.1).

Given the complexity and bidirectional relationship between sleep and drug use during adolescence, each pathway will be addressed separately based on

M. L. Andersen (✉) · G. N. Pires · S. Tufik
Departamento de Psicobiologia, Universidade Federal de São Paulo, Rua Napoleão de Barros, 925, Vila Clementino, SP-04024-002, São Paulo, Brazil
e-mail: ml.andersen12@gmail.com

D. De Micheli et al. (eds.), *Drug Abuse in Adolescence*,
DOI 10.1007/978-3-319-17795-3_9

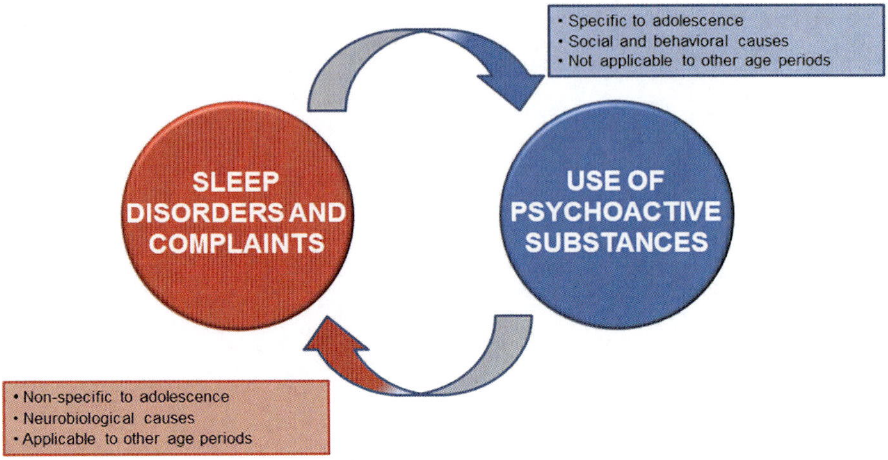

Fig. 9.1 The relationship between sleep-related characteristics during adolescence and drug abuse. These topics are bidirectionally correlated; therefore, both can play the role of cause or effect. Note the differences in how each pathway acts as a causal factor

epidemiological data and information regarding sleep and adolescent chronobiological patterns. To assist readers with understanding this chapter, Table 9.1 defines some variables related to sleep.

Sleep Patterns During Adolescence

Adolescence is a very important period in life, encompassing changes across anatomical, physiological, behavioral, and social domains. With regard to sleep and its associated variables, this period is characterized by several ontogenetic changes.

Table 9.1 Definitions of sleep-related variables

Variable	Definition
Sleep architecture	Organization and transition between sleep stages throughout the night
Sleep latency	Time elapsed between the start of the polysomnographic record (or lying in bed) and sleep onset
REM sleep latency	Time between the start of sleep and REM sleep onset
Total sleep time	Effective sleep time during the polysomnographic record
Sleep efficiency	Percentage of total sleep time relative to the time recorded

From birth to old age, sleep patterns are dynamic and subject to constant change. Certain sleep-related variables show continuous increases or decreases throughout life. This is the case for total sleep time, which is approximately 16 h among newborns but might be only 5 h among the elderly. In this case, adolescence is a period of intermediate values between childhood and adulthood. However, other variables change in a nonlinear way throughout life, through which the striking patterns characteristic of adolescence can best be observed. This is the case for chronobiological preference. A morning type preference is common during both early childhood and late adulthood, whereas an evening type preference is typical during adolescence. Moreover, most of the sleep-related variables exhibit certain idiosyncrasies during adolescence. Thus, adolescence is a unique period of ontogenetic development, especially considering the sleep architecture, the prevalence of sleep disorders, and chronobiological preference.

Sleep Architecture and Total Sleep Time

Sleep is structured in the same way for all ages since the maturation of circadian control, which occurs at approximately 3 months old [3]. In general, sleep is a functional, cyclic, and reversible state. Sleep is primarily divided into two cyclic stages: rapid eye movement (REM) and non-REM (NREM) sleep. Each sleep cycle lasts between 90 and 110 min and consists of an NREM sleep episode followed by an REM sleep episode. Thus, four to six NREM-REM cycles occur in one night of sleep. The NREM stage is further divided into three stages: N1, N2, and N3. These stages show a progressive decrease in various physiological functions as well as reduced EEG activity. Table 9.2 presents more detailed information about each sleep stage [4, 5].

Table 9.2 Stages of the wake-sleep cycle

Sleep stage		Characteristics
Awake		High brain activity with high frequency and low amplitude waves. Predominance of beta waves (above 13 Hz) during attentive wakefulness and alpha waves (8–13 Hz) during relaxed wakefulness
NREM	N1	Transition between sleep and wakefulness. Attenuation of brain electrical activity and the appearance of theta waves (3–7 Hz). Presence of vertex sharp waves
	N2	EEG slowing with a predominance of theta waves. Presence of K complexes and sleep spindles
	N3	Also known as slow wave sleep because of high amplitude and low frequency EEG tracing. Predominance of delta waves (0.5–2 Hz)
REM		Characterized by the presence of tonic events (e.g., muscular hypotonia) and phasic events such as eye movement. Desynchronized brain electrical pattern similar to that of wakefulness, accompanied by sawtooth waves (2–6 Hz)

The first observable changes during adolescence occur with regard to total sleep time. On average, adolescents sleep from 7.5 to 8 h a night, significantly less than the 10 h of sleep required during childhood [6]. As explained above, total sleep time decreases throughout life. In addition to the expected ontogenetic decrease, however, the effect of the combination of eveningness and a morning school schedule might play an important role during adolescence affecting issues such as sleep pressure (more details concerning this relationship will be provided later when chronotype is discussed). Furthermore, studies relying on subjective research methods such as questionnaires tend to underestimate total adolescent sleep time by approximately 0.5 h compared with those using objective methods such as actigraphy [7].

The major change in sleep architecture during adolescence when compared with childhood is a decreased on N3 sleep stage, which can be seen both measuring N3 sleep time or percentage of this stage [8, 9]. From the first months until the second decade of life, N3 stage may decrease about 40% [8]. A similar decrease in REM sleep can be noted, however, only in terms of time on N3 stage. When considered percentage, the proportion of N3 remains stable during all this period, due to an overall decrease in total sleep time [3]. In addition, adolescents have more N2 sleep and a reduced REM sleep onset latency. No differences exist regarding sleep latency and efficiency compared with childhood [3]. Figure 9.2 illustrates the ontogeny of sleep-related variables, comparing adolescents with other age groups.

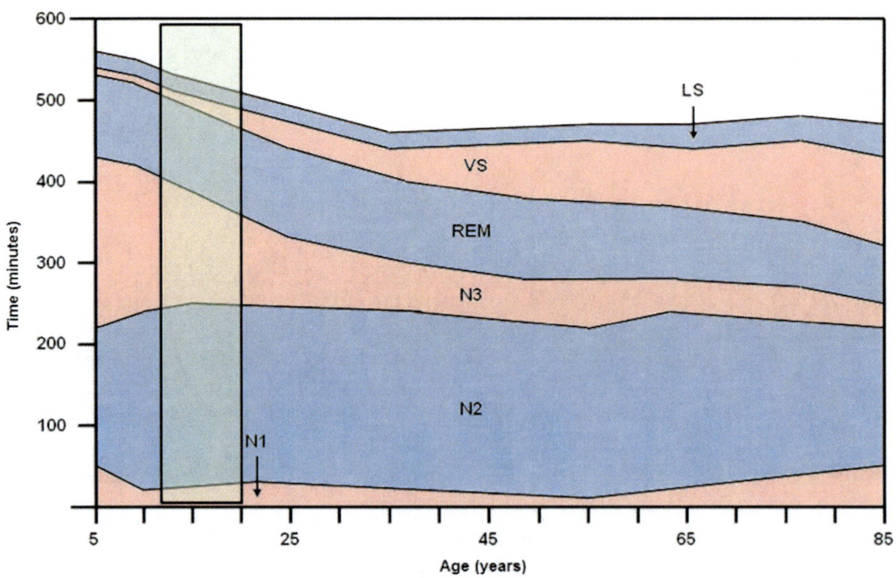

Fig. 9.2 Sleep ontogeny. This figure shows the ontogeny of several sleep-related variables with a particular emphasis on adolescence. *WASO* wake after sleep onset, *SL* sleep latency. (Adapted from Roffwarg et al. [64] and Ohayon et al. [9])

Chronotypes

Chronotypes are profiles used to address the chronobiology associated with circadian preference. According to the currently valid definitions proposed by Horne and Ostberg, the population can be divided into three types with regard to preferences for sleep and wakefulness: indifferent, morning, and evening types. Morning and evening types can be subdivided into moderate and extreme [10].

Morning individuals are those who prefer to wake up early, prefer to perform routine activities at the beginning of the day, and that usually go to sleep early. These individuals have biological rhythms that are shifted ahead of the population mean. On the other hand, evening individuals are those who prefer to wake up late, reach the pinnacle of performance during the second half of the day, and prefer to go to sleep late. These individuals exhibit biological rhythms that lag behind the general population. Finally, indifferent individuals are those who prefer to sleep and wake on an average time, who do not have clear chronobiological preferences, and who are able to more easily adapt to small changes in schedules [11].

The evening chronotype characterizes adolescence. This profile, which begins at approximately 13 years old and peaks at 20 years old, is observed in 11 % of all adolescents [8, 12]. The proportion of adolescents with an evening chronotype is higher than that observed among both children and adults. It is undeniable that social, behavioral, and family factors contribute to eveningness among adolescents. In this sense, several factors influence sleep such as extracurricular activities, homework, night jobs and extended shifts, reduced parental control over sleep schedules and recreation, and leisure and entertainment activities [3]. However, recent studies have shown that the main cause of eveningness during adolescence relies on changes in endogenous rhythmicity, affecting both circadian and homeostatic processes (Fig. 9.3). Thus, the characteristic adolescent eveningness is not merely a consequence of extrinsic behavioral and social features, but also result of biological/intrinsic adaptation and changes during this period. The main biological/intrinsic causes of the higher prevalence of eveningness likely include the following [13]:

- a lower sensitivity to sleep pressure;
- a circadian period extended to approximately 24.3 h; and
- an increased effect of light as a rhythm synchronizer to delay sleep phases.

Sleep Disorders

With regard to the prevalence of sleep disorders, adolescence is clearly a period of transition. During this stage of life, the prevalence of childhood sleep disorders such as parasomnias in general, sleep enuresis (bedwetting), sleepwalking (somnambulism), and bruxism tends to lessen, while adulthood disorders such as insomnia start to appear [8, 14]. Respiratory sleep disorders (such as obstructive sleep patterns) show no ontogenetic pattern, and the causes of these disorders during adolescence can be similar to those observed in adults and children [15].

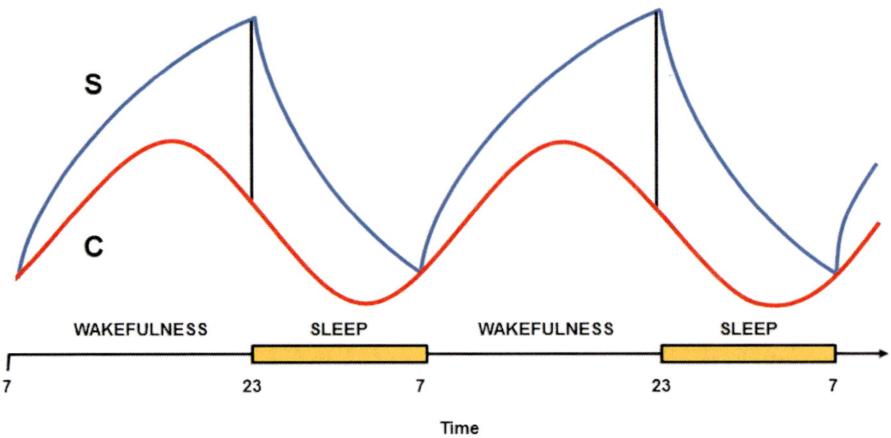

Fig. 9.3 Circadian and homeostatic control of sleep. Sleep is controlled by two distinct processes: process C (*blue line*), which is related to circadian influence and the effect of zeitgebers (synchronizers) on sleep; and process S (*red line*), which is related to the metabolic and homeostatic control of sleep and associated with sleep pressure. Sleep occurs when the influence of both processes is sufficiently large. In contrast, sleep ends when the influence of both processes is low enough to end sleep. Theoretically, the perception of process C is altered, the period of circadian oscillation is extended, and a delay occurs in the process S curve among adolescents. This hypothesis would explain a tendency toward eveningness during this timeperiod. (Adapted from Borbély [65] and Borbély and Achermann [66])

Because of the high prevalence of eveningness, the assessment of rhythm disorders (especially delayed sleep phase syndrome) is important during adolescence. The evening chronotype itself is not a sleep disorder. Therefore, the characterization of sleep phase delay syndrome is associated with extreme eveningness and the inability to synchronize with social routine, which is preferably confirmed using actigraphy and a sleep diary [16]. Prevalence estimates of this syndrome in the general population range from 0.3 to 3.1 %, although it is sometimes confused with the evening chronotype [17]. Sleep phase delay syndrome is estimated to occur in 0.4 % of adolescents [18].

Other sleep disorders, such as movement disorders and narcolepsy, have no marked relationship with adolescence, although they are sporadically observed [4].

Epidemiology of Drug Abuse and Sleep Complaints Among Adolescents

From a public health perspective, the relationship between drug use and sleep complaints/disorders during adolescence is of major importance, mainly due to the high prevalence of these conditions, either assessed independently or in concomitance. Prevalence estimates of sleep-related problems among adolescents vary between 5 and 91 % [14, 19–21]. The significant variability in the data is primarily because of methodological differences in the assessment of these variables. In general, studies of sleep disorders (e.g., insomnia or obstructive sleep apnea) have lower rates than those assessing nonspecific sleep complaints. Studies evaluating specific sleep complaints often obtain median values when they are based on structured questionnaires; moreover, those based on subjective reports of sleep time tend to overestimate the results compared with those based on objective measures. Another possible source of variation in the data concerns who answers the questionnaire; most of the time, the adolescent performs this task, whereas other studies have administered questionnaires to parents. Finally, and in a limiting way, few studies focused exclusively on adolescents who underwent polysomnography (the gold standard sleep assessment). Approximately 51 % of adolescents have trouble sleeping at least once during the week; however, only 0.7 % are diagnosed with idiopathic insomnia [14]. With regard to circadian preference, approximately 11 % of adolescents show the evening chronotype [12]. Although these data are averages, they illustrate the prevalence of sleep-related complaints, which are closely related to low quality sleep and sleep deprivation.

Like the high prevalence of sleep disorders and sleep-related complaints among the general adolescent population, drug abuse among adolescents whose sleep patterns are inadequate is common. Johnson and Breslau first observed this relationship in a sample of 13,831 US adolescents [19]. These authors demonstrated that the percentage of adolescents with sleep complaints was higher among those who used psychotropic substances; furthermore, the percentages of sleep-related complaints among adolescents who used tobacco daily and non-users were 17.5 and 6.4 %, respectively. The same pattern was observed with regard to alcohol consumption (4.2 % among non-users and 10.7 % among those who use alcohol at least once a week) and illicit drugs in general (6.4 % among non-users and 19.5 % among those who use at least once a week). In addition, a logistic regression analysis indicated that the use of tobacco, alcohol, or other drugs increases the risk of sleep-related complaints by approximately three times. Despite the large sample size and the importance of the study in question, sleep was addressed only superficially using a single question regarding the frequency of sleep-related problems over the last 6 months.

Roane and Taylor [22] addressed insomnia during adolescence as a potential risk factor for substance use during early adulthood. That prospective study examined 4494 individuals in the US who were between 12 and 18 years old at baseline and between 18 and 25 years old at the follow-up assessment. The authors found that 57.8 % of adolescents with insomnia reported difficulties falling or staying asleep on a frequent or daily basis. At baseline, the uses of alcohol, cannabis, and other illicit drugs among adolescents with insomnia were 32.7, 18.2, and 7.3 %, respectively, whereas those for adolescents without insomnia were 22, 11.5, and 4.1 %, respectively. Over the long term, insomnia during adolescence did not result in a higher incidence of the use of these substances into early adulthood. The data indicated that the relationship between sleep and drug use does not seem to have a long-term effect; rather, one factor reflects the other synchronously. In a similar analysis specific to tobacco, however, Patten et al. [23] examined 7960 individuals and reported that smoking increases by the risk of complaints related to long-term sleep by approximately three.

Despite having a smaller sample than the previously presented studies (703 South African adolescents between 13 and 20 years old), the study conducted by Fakier and Wild [24] deserves attention for its greater care with regard to data acquisition. In that case, sleep disorders were assessed using the Child Behavior Checklist. In turn, drugs of abuse were assessed in detail across nine categories: tobacco, alcohol, methamphetamine, cannabis, methaqualone, inhalants, crack/cocaine, ecstasy, or other. The prevalence was higher among adolescents with sleep complaints with regard to the use of all substances analyzed (except for methaqualone). After controlling for gender, age, and especially learning difficulties, however, sleep problems were considered an independent risk factor for uses of tobacco, cannabis, and inhalants.

Finally, recent studies have shown that adolescent drug abuse can be stratified by sleep time, being more prevalent among individuals with the evening chronotype than those with the morning chronotype. This relationship has been observed with regard to the use of various substances, always indicating greater use among evening adolescents. The evaluated substances included tobacco, alcohol, caffeine, and cannabis [12, 25, 26]. However, the cause for this relationship is not entirely clear. Sleep deprivation caused by the disparity between chronotype and social demands as well as the behavioral characteristics of adolescence might contribute to these results [1, 27].

Previous studies have suggested the presence of a causal relationship between sleep disorders and drug use. Although statistical data suggests a causal relationship between these phenomena, the direction of this relationship in practical terms is unclear. Thus, predicting which relationship component is the cause and which is the effect is difficult. In fact, it is wise to consider this association as bidirectional (see Fig. 9.1), and the causal relationship should only be assessed in prospective and long-term contexts.

Effects of Drug Abuse on Sleep

The effects of drug abuse on sleep patterns, sleep-related complaints, and sleep disorders occur through their mechanisms of action and pharmacological properties. The action mechanisms of drugs of abuse are the same for any age group; however, small particularities differentiate adolescents from adults and children. These differences are primarily related to neurochemical characteristics and the composition of neurotransmitter systems [28]. One example is the increase in dopaminergic activity in the striatum during adolescence [29] that might influence the establishment of addiction. Therefore, the effects of drug abuse on sleep are not specific to adolescence; although changes exist in terms of the neurochemistry in comparison to other age groups, there are few practical differences. Although adolescents are susceptible to the effects of drug use on their sleep patterns, similar results are observed during adulthood.

The major acute effects of drugs of abuse on sleep are due to their substance characteristics or classification. The acute effects of stimulants include fragmented and non-restorative sleep as well as decreased total sleep time. Depressants such as alcohol, benzodiazepines, and opioids primarily cause increased sleepiness and decreased sleep latency as well as sleep fragmentation during the second half of the night, affecting mainly REM sleep [1, 30]. In addition, chronic sleep effects are commonly observed for most drugs of abuse. These effects include increased sleep latency, reduced total sleep time, increased nocturnal awakenings, and decreased REM sleep time and slow-wave sleep [1]. With regard to abstinence periods, an increase in sleep latency and decreases in slow-wave sleep and total sleep time are observed. In general, maintaining abstinence leads to the return of a normal sleep patterns [1, 30]. Importantly, most studies on drug abuse and sleep have been performed during periods of abstinence because of the greater difficulty associated with performing studies on current users. Specific effects of certain drugs of abuse are addressed below.

Alcohol

Although alcohol is commonly used to induce sleep, its acute effects are deleterious, especially with regard to the fragmentation of sleep. Being a drug with GABAergic agonist action, alcohol induces decreased sleep latency and increased slow-wave sleep. For the same reasons, REM sleep is inhibited, and fragmented sleep is observed as a consequence. Furthermore, depending on the dose ingested and the individual's metabolic rate, REM sleep rebound is often observed, while plasma levels of alcohol diminish [31, 32]. During the abstinence period, complaints of insomnia associated with polysomnographic findings of decreased total sleep time, slow-wave sleep, and REM sleep rebound are common. Finally, the acute use of alcohol increases snoring and may worse a previous case of obstructive sleep apnea, given its action on muscular tone [31, 32].

Tobacco

Despite its widespread use both among the general population and adolescents, few studies have addressed the effects of tobacco on sleep. Tobacco might lead to non-restorative sleep and difficulties initiating sleep [31]. Studies using chewing gum or nicotine patches have described decreases in slow-wave sleep and sleep efficiency as well as increased latencies to initiate sleep [33, 34]. The major effects regarding abstinence are fragmentation of sleep, leading to a superficial sleep and, thereby resulting in excessive daytime sleepiness [31]. Regarding sleep disorders, smoking is also related to bruxism and snoring [35–37]. Concerning obstructive sleep apnea, tobacco use is related to hemoglobin desaturation and to an increased arousal rate, but not to the quantity of respiratory events (more specifically, to the apnea-hypopnea index), nor does it constitute an independent risk factor for the development of this syndrome [38, 39].

Opioids

In addition to their use as drugs of abuse, opioids have high clinical value, mainly when used as analgesics. Although opioids are depressants, their hypnogenic properties are not as striking as those of benzodiazepines or alcohol. Although patients in chronic pain sleep better when using opioids (particularly morphine), this effect is primarily due to the pain relief that enables sleep, since these substances have relatively weak direct hypnogenic mechanisms. The use of opioids can lead to transitory decreases in total sleep time and REM sleep inhibition [31]. Regarding abstinence, problems initiating and maintaining sleep as well as low sleep quantity and quality have been reported during treatment with methadone [40].

Cocaine

The major effects of acute cocaine use on sleep include increased sleep latency, reduced total sleep time, and REM sleep suppression. In addition to these polysomnographic findings, users commonly complain of difficulties falling asleep associated with euphoric feelings and exacerbated alertness. In general, these effects are likely because of the increased synaptic availability of dopamine [30, 31].

All effects of acute abstinence shorten sleep time. Thus, decreases in sleep efficiency and total sleep time as well as increased sleep latency have been observed. Moreover, decreases in REM sleep latency and increases in the prevalence of this stage have been reported. These symptoms tend to worsen as abstinence becomes subacute (approximately 10 days), with deteriorations in sleep quality as evidenced by decreased total sleep time and increased sleep latency [30, 41]. This sleep impairment is maintained for approximately 3 weeks after cessation and tends to normalize with maintained abstinence [1, 42].

Cannabis

There is much debate concerning the effects of the use of cannabis and endocannabinoids with regard to sleep patterns because of the methodological limitations of the studies conducted on this subject (i.e., low sample sizes, variability between dosage and the route of administration, and its concomitant use with other substances) [30]. Nevertheless, certain common effects can be listed. Cannabis use is associated with decreases in sleep latency and subjective reports of the increase ease of falling asleep; however, the opposite is observed when using high doses of

Δ-9-tetra-hidrocannabinol. In addition, polysomnographic studies show increases in slow-wave sleep and decreases in REM sleep time [30]. Both the effects concerning sleep induction and slow-wave sleep depend on tolerance [30]. Regarding abstinence, 2 weeks after cessation, decreases in total sleep time and sleep efficiency have been observed [43]. Abstinence effects on REM sleep remain contradictory; however, they are likely related to increased reports of dreams [30].

Ecstasy

The acute use of ecstasy (3,4-methylenedioxymethamphetamine, MDMA) or its analogue 3.4-methylenedioxy-N-etilanfetamine1 (MDE) affects sleep quality, supporting the reports of non-restorative sleep up to 48 h after their use [44]. In addition, increases in wakefulness and a almost total REM sleep suppression have been reported [45]. These effects are associated with the dopaminergic and serotonergic actions of the substances in question [30].

The results concerning the sleep effects of the continued use of ecstasy should be interpreted carefully because of several methodological biases (i.e., the concomitant use of other drugs of abuse and the use of improperly controlled methodologies) [30]. The few studies that address this relationship have reported increases of stage N2 sleep (a finding with little practical relevance) as well as increased sleep efficiency and slow-wave sleep time [46, 47].

Sleep During Adolescence as a Predisposing Factor to Drug Abuse

Although the effect of drug abuse on adolescent sleep patterns is a nonspecific association that is observed among other age groups, the inverse relationship is specific and unique. In fact, the sleep patterns of adolescents influence drug abuse in a way that is not common among other ages. Thus, although sleep-related variables influence drug abuse among other age groups, their causes must differ from those discussed in this chapter.

Sleep during adolescence differs from that observed during adulthood. In general, these differences concern sleep architecture, development, the prevalence of sleep disorders, and (most commonly) day and chronotype preferences (see above). Furthermore, the behavioral phenotype during adolescence is typical, and it should certainly affect this relationship. However, many questions can be raised regarding how the behavior and the chronotype interact to generate a greater predisposition to drug abuse. The following section discusses how adolescent sleep can result in the increased use of addictive substances, the importance of the behavioral phenotype on this relationship, the roles of family and social activities, and drug use as a form of self-medication among sleep deprived individuals.

Chronotype and Drug Abuse

This chapter discussed and established both the high prevalence of the evening chronotype and the pattern of drug abuse above. As previously mentioned, the prevalence of drug abuse among adolescents with the evening chronotype is higher than that among adolescents with the morning chronotype [12, 25, 26]. According to Giannotti et al. [12], evening adolescents consume more caffeine, hypnogenic substances, and drugs of abuse in general. Gau et al. [25] demonstrated that the uses of tobacco and alcohol are significantly higher among adolescents with the evening chronotype. Other studies have corroborated these findings [48, 49]. When males are analyzed specifically, the prevalence of drug abuse, especially tobacco, is higher among those with the evening chronotype [50].

Based on these results, chronotype and drug abuse are likely associated with one another in a causal relationship: Specifically, the evening chronotype causes increased rates of drug abuse during adolescence. However, how these topics are related remains unclear. Although these phenomena might be directly related, chronotype and drug use are more likely two extremes of a causal relationship, while certain other factors may play an intermediate role on it. Drugs of abuse are the effect because it is the outcome with greater relevance. In turn, chronotype becomes the primary cause because it is an inherent characteristic of adolescents and (to some extent) independent of environmental factors [1, 13]. The major factors that mediate this relationship are chronic sleep deprivation, the common behavioral phenotypes of evening adolescents, and self-medication. Figure 9.4 shows how theoretical models of chronotype and drug use interact during adolescence; the aforementioned mediators are discussed below.

Eveningness and Sleep Deprivation

Eveningness is not a sleep disorder, nor does it necessarily imply sleep deprivation or restriction. Rather, it is an adaptable sleep pattern observable in the general population (although it is more prevalent during adolescence). In fact, even individuals with the evening chronotype can perform their activities without major problems and have sufficient sleep quality and quantity when they are able synchronize their social routines to their sleep patterns. Although chronotypes vary among individuals, social routines are not always flexible; thus, this proposed synchronization might be unrealistic.

This synchronization is especially difficult among adolescents. Although these individuals usually develop an evening sleep pattern, their school-related activity schedule usually requires a morning sleep pattern. Adolescents sleep late because of an inherent characteristic of their sleep patterns but must wake up early for school, therefore subjecting themselves to chronic sleep restriction. Adolescent sleep deprivation due to school routine is a common finding of several studies [2, 3, 8, 12, 21, 51, 52]. One possible solution for this condition would be to change their class schedules to the afternoon, thereby allowing adolescents with an

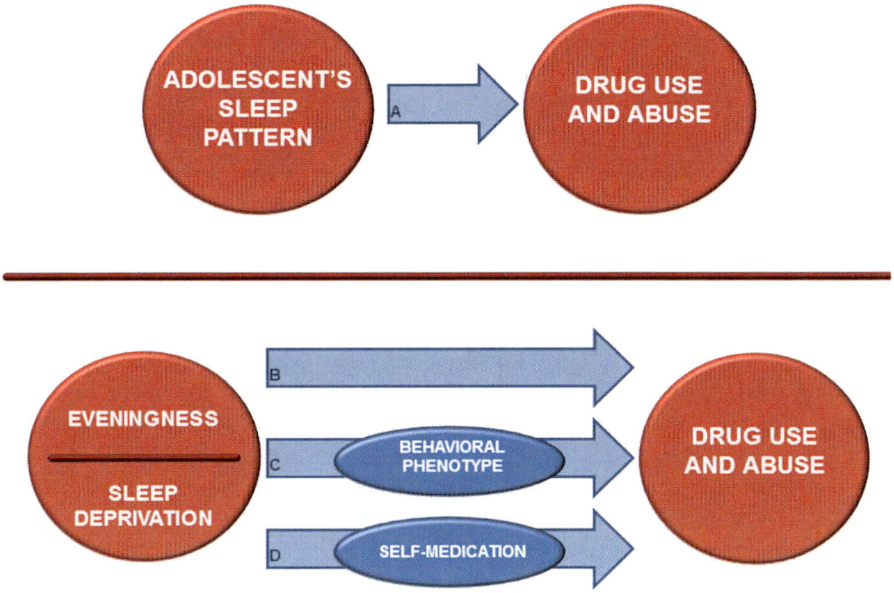

Fig. 9.4 Ways in which sleep patterns during adolescence might predispose individuals to drug abuse. **a** A simple and direct relationship in which the sleep patterns observed during adolescence lead to a greater predisposition to drug abuse. **b**, **c**, and **d** Detailed relationships between the two factors. In these cases, drug use during adolescence is primarily caused by two sleep features characteristic of adolescence: eveningness and sleep deprivation. Both were analyzed together because they are practically inseparable. **b** The adolescent sleep pattern predisposes individuals to drug abuse. In this case, drug abuse is a direct consequence of adolescent sleep patterns and can be considered an integral behavior of the adolescent behavioral phenotype. **c** Behavioral characteristics commonly found in adolescents that mediate the relationship. In this case, sleep patterns predispose individuals to a behavioral profile characterized by externalizing behavioral problems, which can be a risk factor for drug abuse or lead adolescents to engage in risky situations. **d** The relationship mediated by self-medication in which depressants or stimulants are used to induce sleep or circumvent the excessive daytime sleepiness that results from chronic sleep deprivation

evening chronotype to obtain an adequate amount of sleep each night. Although this solution can be extensively discussed from a theoretical perspective, no practical study has addressed this issue. The closest study described an 85-minutes delay to the start of classes [53]. Despite being able to wake up an hour later during school days, students maintained their usual sleep schedule. Although this study did not examine school performance or drug abuse, this change resulted in fewer missed classes and less reports of depression among students.

The sleep restriction induced by morning school activities becomes more evident when sleep time during the school days and weekends/holidays are compared. Giannotti et al. [12] studied 16- to 18-year-olds and reported that the sleep times of morning individuals differed by only 20 min between school days and weekend days, sleeping on average 480 min in the first case and 500 min in the second. However, the sleep times of evening adolescents differed by 90 min; they slept approximately 440 min during school days and 530 min during weekends.

Several reports have indicated that the sleep deprivation caused by school routine causes several problems related to adolescent behavior and academic performance. These problems include attention deficits, higher levels of stress, a higher incidence of low grades, aggressive and delinquent behaviors, and suicidal ideation among others [12, 21, 25, 26]. In addition, reduced sleep times in adolescents have been related to increased substance use including caffeine, cannabis, tobacco, and alcohol [12, 21, 22, 25]. In fact, increases in drug abuse are likely the result of the misalignment between students' biological synchronization and their social and school requirements [27]. Roane and Taylor [22] indicated that adolescent insomnia, which is a consequence of the disparity between social demands and the sleep phase delay observed among evening individuals, predisposes these individuals to drug abuse in general. Tynjälä et al. [54] studied another result of adolescent sleep deprivation and reported similar results using self-reported subjective fatigue on school days. Noland et al. [21] reported that an inadequate amount of sleep (defined as a total sleep time briefer than 9 h per night) is related to the use of several drugs of abuse. In that study, 6% of adolescents who met the criteria for insufficient sleep reported using hypnogenic drugs, 5.7% reported smoking before bedtime, and 2.9% reported consuming alcohol at night to promote sleep. However, this sample [21] was subject to bias because 91.9% of the students assessed had insufficient sleep.

A critical analysis of Noland et al. [21] reveals an important feature of adolescent sleep. The prevalence of adolescents who report insufficient sleep or sleep deprivation is high. Thus, differentiating whether behavioral effects (including drug abuse) are due to an inherent characteristic of the evening chronotype or whether they are associated with sleep deprivation becomes difficult. In practical terms, however, the difficulty associated with differentiating these effects among adolescents does not reduce the applicability of the results because eveningness and sleep deprivation are concomitant and virtually inseparable.

Behavior as a Mediator Between Sleep and Drug Abuse

The eveningness-sleep deprivation binomial is a causal factor of drug use during adolescence, as the data presented previously clearly demonstrate. Because of the concomitant presentation of each part of this relationship, their independent effects are difficult to assess; therefore, it is more sensible to approach them as a single entity (the aforementioned binomial). In addition to a greater predisposition to drug use, other consequences include behavioral changes, encompassing both internalizing and externalizing problems. This finding begs the question of whether adolescents' increased use of psychoactive substances is a direct function of sleep deprivation or the result of a behavioral spectrum generated by a lack of sleep (Fig. 9.4). As previously mentioned, several studies have shown the direct effect of sleep patterns on the use of drugs. However, drug abuse might be part of a complex behavioral phenotype caused by sleep deprivation and eveningness. By analyzing this relationship thoroughly, chronic sleep deprivation and eveningness alone were determined to be responsible for many neurobehavioral consequences in adolescents [51, 55,

56]. Other studies have related behavioral problems to the higher prevalence of drug use and abuse [57, 58].

The importance of behavior in the relationship between sleep and drug use is clear in the work of Johnson and Breslau [19]. In their primary analysis, the sleep complaints of a sample of adolescents were strongly associated with the use of alcohol, tobacco, and illicit drugs in general. After performing regression analyses controlling for internalizing (e.g., anxiety, fear, and depression) and externalizing behavioral problems (e.g., impulsivity, aggression, and delinquent behavior), however, the association became weaker, demonstrating that behavior affects this relationship. Several additional studies have evaluated the effect of the type of behavioral problems on drug use. Although some studies have found significant associations related to internalizing problems, most results indicate an association between drug use and externalizing problems [59–61]. These findings allow us to speculate that externalizing behavior problems are a risk factor for drug abuse. In this case, this behavioral profile would likely increase adolescents' exposure to risky situations regarding the use of drugs. Externalizing behavioral problems, decreased parental control concerning leisure activities, greater freedoms related to social activities compared with childhood, and the influence of social environment make this hypothesis regarding adolescent behavior plausible [3, 8].

Self-Medication

Self-medication is an alternative way through which sleep complaints and disorders predispose individuals to drug use. When placed in a social and educational context, evening adolescents have difficulties with obtaining enough quality sleep. Thus, the abuse of drugs and medications becomes a common practice to circumvent these problems. In this context, hypnogenic medications, antidepressants and depressants substances are commonly used at night to induce sleep; furthermore, stimulants are used to increase alertness and minimize the excessive daytime sleepiness due to chronic sleep deprivation [1, 23, 62]. However, chronic sleep deficits, excessive daytime sleepiness, and self-medication create a vicious cycle in which one factor exacerbates the next [2]. Despite the absence of acute effects, substances such as alcohol and nicotine can lead to chronic sleep and attention impairments as well as learning difficulties [24]. The use of tobacco, alcohol, and hypnogenic drugs (i.e., sleeping pills, including benzodiazepines and even herbal drugs) is common among self-medicators [21, 62]. In addition, their use is often observed among evening adolescents [21].

Sleep Hygiene

Sleep hygiene concerns numerous habits, behaviors, environmental conditions, and other factors that assist in maintaining good sleep quality [63].

Bootzin and Stevens [2] developed an approach applicable to adolescents during drug abuse cessation. In this case, the authors considered sleep-related complaints to be risk factors for relapse. Thus, the treatment of possible sleep disorders and improvements to sleep quality variables were protective factors for relapse. Recommendations and education measures related to sleep hygiene were applied in combination with meditation and psychotherapeutic techniques. When these interventions were applied, positive effects according to subjective reports of sleep, behavior, and abstinence rates were recorded at a follow-up assessment.

Obviously, the use of sleep hygiene techniques is not the only method of addressing adolescent behavior and drug abuse. Of all the alternatives, however, these techniques are only related to sleep. Therefore, these practices can be useful tools to fight drug abuse during adolescence; they should be taught and encouraged among both adolescents and their parents [21].

Conclusions

Sleep and chronobiology are topics of great importance with regard to drug abuse during adolescence. Because they form a bidirectional relationship, sleep and its associated complaints are intricately associated with patterns of drug use. Although the characteristics of specific substances can affect sleep, poor sleep quality during adolescence might be a causal and predisposing factor for drug abuse. Moreover, the latter pathway of the relationship between sleep during adolescence and drug use can be explained by factors such as eveningness, self-medication, and behavior.

The arguments discussed in this chapter emphasize the importance of the relationship studied here. Despite the relevance of this topic, few practical approaches have considered sleep and its relationship to drug abuse during adolescence. This finding is valid for both directions of the proposed bidirectional relationship: 1) sleep is poorly addressed within prevention and education policies as a factor related to drugs; 2) although adolescent sleep characteristics strongly influence drug use, they are not discussed in the context of the risk factors for use, nor are treated to prevent drug use or relapse.

Thus, in addition to describing the relationship between sleep and drug use during adolescence, this chapter demonstrates the need for this topic to be incorporated into prevention and treatment policies related to drug abuse for adolescents.

Disclosure Gabriel Natan Pires is a current employee of Springer; this position has no relation with the present chapter, nor with any of this author's academic activities.

References

1. Hasler BP, Smith LJ, Cousins JC, et al. Circadian rhythms, sleep, and substance abuse. Sleep Med Rev. 2012;16(1):67–81.

2. Bootzin RR, Stevens SJ. Adolescents, substance abuse, and the treatment of insomnia and daytime sleepiness. Clin Psychol Rev. 2005;25(5):629–44.
3. McLaughlin Crabtree V, Williams NA. Normal sleep in children and adolescents. Child Adolesc Psychiatr Clin N Am. 2009;18(4):799–811.
4. Tufik S. Medicina e Biologia do Sono. São Paulo: Manole; 2008.
5. Iber C, Ancoli-Israel S, Chesson A, et al. Medicine AAoS. The AASM Manual for the scoring of sleep and its associated events: rules, terminology and technical specifications. Westchester: American Academy of Sleep Medicine; 2007.
6. Wolfson AR, Carskadon MA, Acebo C, et al. Evidence for the validity of a sleep habits survey for adolescents. Sleep. 2003;26(2):213–6.
7. Beebe DW, Fallone G, Godiwala N, et al. Feasibility and behavioral effects of an at-home multi-night sleep restriction protocol for adolescents. J Child Psychol Psychiatry. 2008;49(9): 915–23.
8. Colrain IM, Baker FC. Changes in sleep as a function of adolescent development. Neuropsychol Rev. 2011;21(1):5–21.
9. Ohayon MM, Carskadon MA, Guilleminault C, et al. Meta-analysis of quantitative sleep parameters from childhood to old age in healthy individuals: developing normative sleep values across the human lifespan. Sleep. 2004;27(7):1255–73.
10. Horne JA, Ostberg O. A self-assessment questionnaire to determine morningness-eveningness in human circadian rhythms. Int J Chronobiol. 1976;4(2):97–110.
11. Benedito-Silva AA. Cronobiologia do ciclo vigília-sono. In: Tufik S, editor. Medicina e biologia do sono. São Paulo: Manole; 2008, pp. 24–33.
12. Giannotti F, Cortesi F, Sebastiani T, et al. Circadian preference, sleep and daytime behavior in adolescence. J Sleep Res. 2002;11(3):191–9.
13. Hagenauer MH, Perryman JI, Lee TM, et al. Adolescent changes in the homeostatic and circadian regulation of sleep. Dev Neurosci. 2009;31(4):276–84.
14. Ivanenko A, Gururaj BR. Classification and epidemiology of sleep disorders. Child Adolesc Psychiatr Clin N Am. 2009;18(4):839–48.
15. Marcus CL. Obstructive sleep apnea syndrome: differences between children and adults. Sleep. 2000;23(Suppl 4):S140–1.
16. Dagan Y, Eisenstein M. Circadian rhythm sleep disorders: toward a more precise definition and diagnosis. Chronobiol Int. 1999;16(2):213–22.
17. Wyatt JK. Delayed sleep phase syndrome: pathophysiology and treatment options. Sleep. 2004;27(6):1195–203.
18. Ohayon MM, Roberts RE, Zulley J, et al. Prevalence and patterns of problematic sleep among older adolescents. J Am Acad Child Adolesc Psychiatry. 2000;39(12):1549–56.
19. Johnson EO, Breslau N. Sleep problems and substance use in adolescence. Drug Alcohol Depend. 2001;64(1):1–7.
20. National Sleep Foundation Sleep and Teens Task Force. Adolescent sleep needs and patterns: research report and resource guide. Washington: National Sleep Foundation; 2000.
21. Noland H, Price JH, Dake J, et al. Adolescents' sleep behaviors and perceptions of sleep. J Sch Health. 2009;79(5):224–30.
22. Roane BM, Taylor DJ. Adolescent insomnia as a risk factor for early adult depression and substance abuse. Sleep. 2008;31(10):1351–6.
23. Patten CA, Choi WS, Gillin JC, et al. Depressive symptoms and cigarette smoking predict development and persistence of sleep problems in US adolescents. Pediatrics. 2000;106(2):E23.
24. Fakier N, Wild LG. Associations among sleep problems, learning difficulties and substance use in adolescence. J Adolesc. 2011;34(4):717–26.
25. Gau SS, Shang CY, Merikangas KR, et al. Association between morningness-eveningness and behavioral/emotional problems among adolescents. J Biol Rhythms. 2007;22(3):268–74.
26. Susman EJ, Dockray S, Schiefelbein VL, et al. Morningness/eveningness, morning-to-afternoon cortisol ratio, and antisocial behavior problems during puberty. Dev Psychol. 2007;43(4):811–22.
27. Wittmann M, Dinich J, Merrow M, et al. Social jetlag: misalignment of biological and social time. Chronobiol Int. 2006;23(1–2):497–509.

28. Izenwasser S. Differential effects of psychoactive drugs in adolescents and adults. Crit Rev Neurobiol. 2005;17(2):51–67.
29. Noisin EL, Thomas WE. Ontogeny of dopaminergic function in the rat midbrain tegmentum, corpus striatum and frontal cortex. Brain Res. 1988;469(1–2):241–52.
30. Schierenbeck T, Riemann D, Berger M, et al. Effect of illicit recreational drugs upon sleep: cocaine, ecstasy and marijuana. Sleep Med Rev. 2008;12(5):381–9.
31. Obermeyer WH, Benca RM. Effects of drugs on sleep. Neurol Clin. 1996;14(4):827–40.
32. Brower KJ. Alcohol's effects on sleep in alcoholics. Alcohol Res Health. 2001;25(2):110–25.
33. Gothe B, Strohl KP, Levin S, et al. Nicotine: a different approach to treatment of obstructive sleep apnea. Chest. 1985;87(1):11–7.
34. Davila DG, Hurt RD, Offord KP, et al. Acute effects of transdermal nicotine on sleep architecture, snoring, and sleep-disordered breathing in nonsmokers. Am J Respir Crit Care Med. 1994;150(2):469–74.
35. Balaguer C, Palou A, Alonso-Fernández A. Smoking and sleep disorders. Arch Bronconeumol. 2009;45(9):449–58.
36. Rintakoski K, Ahlberg J, Hublin C, et al. Bruxism is associated with nicotine dependence: a nationwide Finnish twin cohort study. Nicotine Tob Res. 2010;12(12):1254–60.
37. Franklin KA, Gíslason T, Omenaas E, et al. The influence of active and passive smoking on habitual snoring. Am J Respir Crit Care Med. 2004;170(7):799–803.
38. Hoflstein V. Relationship between smoking and sleep apnea in clinic population. Sleep. 2002;25(5):519–24.
39. Conway SG, Roizenblatt SS, Palombini L, et al. Effect of smoking habits on sleep. Braz J Med Biol Res. 2008;41(8):722–7.
40. Oyefeso A, Sedgwick P, Ghodse H. Subjective sleep-wake parameters in treatment-seeking opiate addicts. Drug Alcohol Depend. 1997;48(1):9–16.
41. Pace-Schott EF, Stickgold R, Muzur A, et al. Sleep quality deteriorates over a binge–abstinence cycle in chronic smoked cocaine users. Psychopharmacology (Berl). 2005;179(4):873–83.
42. Morgan PT, Pace-Schott E, Pittman B, et al. Normalizing effects of modafinil on sleep in chronic cocaine users. Am J Psychiatry. 2010;167(3):331–40.
43. Bolla KI, Lesage SR, Gamaldo CE, et al. Polysomnogram changes in marijuana users who report sleep disorders during prior abstinence. Sleep Med. 2010;11(9):882–9.
44. Huxster JK, Pirona A, Morgan MJ. The sub-acute effects of recreational ecstasy (MDMA) use: a controlled study in humans. J Psychopharmacol. 2006;20(2):281–90.
45. Gouzoulis E, Steiger A, Ensslin M, et al. Sleep EEG effects of 3,4-methylenedioxyethamphetamine (MDE; "eve") in healthy volunteers. Biol Psychiatry. 1992;32(12):1108–17.
46. McCann UD, Ricaurte GA. Effects of (+/-) 3,4-methylenedioxymethamphetamine (MDMA) on sleep and circadian rhythms. Sci World J. 2007;7:231–8.
47. Ricaurte GA, McCann UD. Experimental studies on 3,4-methylenedioxymethamphetamine (MDA, "ecstasy") and its potential to damage brain serotonin neurons. Neurotox Res. 2001;3(1):85–99.
48. Adan A. Chronotype and personality factors in the daily consumption of alcohol and psychostimulants. Addiction. 1994;89(4):455–62.
49. Randler C. Differences between smokers and nonsmokers in morningness–eveningness. Social Behav Personal. 2008;36:8.
50. Negriff S, Dorn LD, Pabst SR, et al. Morningness/eveningness, pubertal timing, and substance use in adolescent girls. Psychiatry Res. 2011;185(3):408–13.
51. Dahl RE, Lewin DS. Pathways to adolescent health sleep regulation and behavior. J Adolesc Health. 2002;31(6 Suppl):175–84.
52. Brandalize M, Pereira RF, Leite N, et al. Effect of morning school schedule on sleep and anthropometric variables in adolescents: a follow-up study. Chronobiol Int. 2011;28(9):779–85.
53. Wahlstrom K. Changing times: findings from the first longitudinal study of later high school start times. NASSP Bulletin. 2002;86(633):3–21.

54. Tynjälä J, Kannas L, Levälahti E. Perceived tiredness among adolescents and its association with sleep habits and use of psychoactive substances. J Sleep Res. 1997;6(3):189–98.
55. Dagys N, McGlinchey EL, Talbot LS, et al. Double trouble? The effects of sleep deprivation and chronotype on adolescent affect. J Child Psychol Psychiatry. 2012;53(6):660–7.
56. Caci H, Mattei V, Baylé FJ, et al. Impulsivity but not venturesomeness is related to morningness. Psychiatry Res. 2005;134(3):259–65.
57. Beyers JM, Toumbourou JW, Catalano RF, et al. A cross-national comparison of risk and protective factors for adolescent substance use: the United States and Australia. J Adolesc Health. 2004;35(1):3–16.
58. Riala K, Ilomäki E, Hakko H, et al. Is the severity of adolescent conduct disorder associated with the level of nicotine dependence? Eur Child Adolesc Psychiatry. 2011;20(8):393–9.
59. Griffith-Lendering MF, Huijbregts SC, Mooijaart A, et al. Cannabis use and development of externalizing and internalizing behavior problems in early adolescence: a TRAILS study. Drug Alcohol Depend. 2011;116(1–3):11–7.
60. Dawson DA, Goldstein RB, Moss HB, et al. Gender differences in the relationship of internalizing and externalizing psychopathology to alcohol dependence: likelihood, expression and course. Drug Alcohol Depend. 2010;112(1–2):9–17.
61. Katz LY, Cox BJ, Clara IP, et al. Substance abuse versus dependence and the structure of common mental disorders. Compr Psychiatry. 2011;52(6):638–43.
62. Stein MD, Friedmann PD. Disturbed sleep and its relationship to alcohol use. Subst Abus. 2005;26(1):1–13.
63. Stepanski EJ, Wyatt JK. Use of sleep hygiene in the treatment of insomnia. Sleep Med Rev. 2003;7(3):215–25.
64. Roffwarg HP, Muzio JN, Dement WC. Ontogenetic development of the human sleep-dream cycle. Science. 1966;152(3722):604–19.
65. Borbély AA. The S-deficiency hypothesis of depression and the two-process model of sleep regulation. Pharmacopsychiatry. 1987;20(1):23–9.
66. Borbély AA, Achermann P. Sleep homeostasis and models of sleep regulation. J Biol Rhythms. 1999;14(6):557–68.

Gabriel Natan Pires BS in Biomedicine, Federal University of Health Sciences of Porto Alegre. Graduate student, Department of Psychobiology, Federal University of São Paulo (UNIFESP).

Monica Levy Andersen BS in Biomedicine, São Paulo State University (Universidade Estadual Paulista Júlio de Mesquita Filho—UNESP). MS and PhD in Psychobiology, Federal University of São Paulo (UNIFESP). Associate Professor of Sleep Medicine and Biology, Department of Psychobiology, UNIFESP. Visiting Associate Professor, Yerkes National Primate Research Center—Emory University, USA.

Sergio Tufik MD, School of Medical Sciences of the Holy House of Mercy of São Paulo (Faculdade de Ciências Médicas da Santa Casa de São Paulo—FCMSC-SP). PhD in Psychopharmacology, Federal University of São Paulo/School of Medicine of São Paulo (UNIFESP/EPM). Full Professor of Sleep Medicine and Biology, Department of Psychobiology, UNIFESP/EPM.

Chapter 10
Neurological and Cognitive Changes Resulting from Chronic Drug Use

Rafaela Larsen Ribeiro, Shirley Yamaguchi, Silvana Pochetto Cracasso, Fernanda Peralta and Elton Rezende

Introduction

For some people, drug use in adolescence is part of the normal process of searching for novel experiences, selfness and maturity that characterize this phase of life. It is also known that most young people who experiment with drugs do not become dependent on them. However, dependence is only one of the risks that young people face when using these substances. There are other extremely serious risks, such as automobile accidents, trauma, drowning, unsafe sexual practices, exposure to sexually transmitted diseases, unwanted pregnancy, and several other risk behaviors, associated with adolescent drug use. Moreover, because transformation and maturation of the central nervous system (CNS) occurs during adolescence, the use of drugs during this period may considerably impair such development, causing damage to the intellectual, emotional and social potential of the youth. This issue must be handled with great care to avoid exaggeration or "demonization" of simple experimentation while at the same time maintaining awareness of the possibility

R. L. Ribeiro (✉)
Department of Psychobiology, Federal University of São Paulo (UNIFESP), São Paulo, Brazil
e-mail: profarafaelaribeiro@gmail.com

S. Yamaguchi
Institute of Psychiatry, School of Medicine, University of São Paulo (USP), São Paulo, Brazil

S. P. Cracasso
Federal University of São Paulo (UNIFESP), São Paulo, Brazil
e-mail: silpepsi@gmail.com

F. Peralta
School of Medicine, University of São Paulo (USP), São Paulo, Brazil

E. Rezende
Federal University of Minas Gerais, Belo Horizonte, Brazil

Federal University of São Paulo (UNIFESP) and University of São Paulo (USP), São Paulo, Brazil

D. De Micheli et al. (eds.), *Drug Abuse in Adolescence*,
DOI 10.1007/978-3-319-17795-3_10

of abuse or dependence. Thus, the family and school are important institutions that should be alert during the development of young people, providing guidance, support and robust emotional and intellectual references. When in doubt or when there is evidence of any behavior that might indicate psychiatric disorders and/or drug use, an evaluation by a psychiatrist becomes necessary [1].

Cognitive Aspects During Adolescence and Neural Mechanisms of Decision-Making

Adolescence is defined by the World Health Organization (WHO) as the period from ages 10 to 19. It represents the transition to adulthood, during which the brain is in a dynamic and unique stage of development. This stage is characterized by an increased demand for novelty and increased risk behavior, which implies a greater risk of initiating psychotropic substance use [2].

Several risk factors are associated with substance use among youth. Studies on neurodevelopment suggest that the brains of adolescents may be more vulnerable to the effects of drugs and related substances than are the brains of adults.

At this stage, important biopsychosocial changes usually occur. These psychological, socio-cultural and neurological aspects have been extensively studied in attempts to understand the association between childhood/adolescence and higher susceptibility to experimentation and the use of drugs [3].

Research based on functional magnetic resonance imaging has shown that new synaptic connections are extensively formed in the adolescent brain, comparable to those that are formed during the immediate postnatal period. During adolescence, the major changes in synaptic connections occur in the prefrontal cortex, a region that coordinates "executive" thinking, the ability to use logic, make decisions and evaluate possible risks [2].

According to Bechara et al. [4], changes in the prefrontal cortex of alcoholics tend to compromise the process of decision- making, inducing the individual to choose the most attractive options with respect to immediate gains (such as, for example, the act of substance consumption) rather than behaviors based on an analysis of the future consequences of those actions. Changes in the orbitofrontal cortex are observed even after months of alcohol abstinence and may be associated with long-lasting changes in the serotonergic and GABAergic activity in this region. These changes influence decision-making, inhibitory control and consumption behavior (seeking alcohol, for example) that maintains substance dependence [5].

Cunha and Novaes [6] state that neurocognitive changes have direct implications for psychiatric treatment, both in the choice of strategies to be adopted and in the analysis of prognosis. A deficit in executive functions would affect treatment, as individuals experiencing such a deficit would find it difficult to understand and assimilate the basic concepts of therapy, to set and accomplish goals that do not involve an immediate reward and to inhibit inappropriate impulsive responses [7].

The frontal lobes endow humans with the ability to perform tasks that other animals seem unable to carry out. Most importantly, they allow humans to free themselves from the present and to project themselves into the future. We can think about the future, plan it, imagine how it will be and, depending on the decisions we make, make the choices we deem most appropriate to our goals. The frontal lobes permit us to engage in future-directed activity. It is with a remarkable capacity that we constantly use this ability, which then becomes almost automatic. The frontal lobes are also responsible for controlling impulses that might negatively affect our plans for the future, causing us to avoid certain activities and dissuading us from acting in ways that are contradictory to our goals [8].

The frontal lobes do not function optimally during adolescence because at that time they are in the process of maturation and are undergoing a period of intense remodeling. During this phase, the individual learns the behavior patterns applicable to a particular culture in a particular time [8].

Thus, at this stage, it is our responsibility to guide adolescents to make wise decisions and to teach them how to treat others, what to do with their time and how to plan ahead. After an individual reaches maturity, it is no longer possible to extensively remodel or modify the brain. At this time, it is assumed that the individual has completed his or her development and is prepared to function in the adult world. Whether individuals live as adults for 10 or for 100 years, adolescence will always be the period that prepares them to function as mature people [8].

According to studies conducted by the American Academy of Pediatrics (AAP) [9], the cultural myths and symbols used in alcohol advertisements effectively influence the consumption of this substance by adolescents. For a developing mind like that of the adolescent, which is especially questioning and plastic, the paradox represented by a society's expressed opinion and its lack of strong law enforcement presents an ideal cultural environment for experimentation with alcohol and other drugs, thereby contributing to the early exposure of young people to abusive consumption.

However, what does plasticity mean? Neural plasticity is the brain's ability to develop new synaptic connections among neurons based on the individual's experience and behavior [10]. With each new learned behavior from birth to adulthood, a number of neural connections occur and are fixed in the CNS, contributing to its normal and evolutionary development. Neural plasticity is innate and is essential to the learning process and to the development of the neuropsychological and motor functions of the individual. Thus, it is possible to stimulate the individual through psychotherapy, specific exercises and training; the greater the appropriate stimuli, the better the level of functioning [10].

One may empirically state that humans do not change greatly after the age of 30. At that point in life, our behavior patterns are somehow "set". We have certain likes and dislikes that will not change significantly at this point. We may change, but it is unusual that our lives will radically change from that age onward [8].

During adolescence, the potential for change is enormous. However, this "window" closes at approximately the age of 20. In this phase of life, the brain is extremely malleable. This is a stage when individuals must acquire and retain a large amount of information. Alcohol is highly effective at blocking brain plasticity and may interfere with neural plasticity both in adolescence and in adulthood, leading to temporary loss of memory (memory gaps) [8].

As the amount of ingested substances increases, greater effects on memory are observed; at some point, the person will lose the ability to recall facts, or his or her memories will be faulty and no longer reflect what actually happened.

Undoubtedly, the young brain seems more susceptible than the adult brain to the effects of alcohol, including its effects on memory. During adolescence, the frontal lobes are not fully developed; thus, individuals are not able to make sound long-term decisions. Associated with this there is potentially the harmful use of alcohol, which prevents one from making the right decisions. Alcohol use often leads individuals to make uncertain decisions, most likely due to its effects on the frontal lobes [8].

In adolescents, the frontal lobes are still undergoing maturation and thus are significantly affected by alcohol, impairing decision-making and impulse control. To conclude, we may say that during the second decade of life a reduction in the volume of gray matter occurs in the frontal lobes and that is when individuals become more effective at performing tasks. The complete process is known as "frontalization"; it means that we enter adolescence as entirely emotional beings driven primarily by feelings and leave it as beings capable of using the frontal lobes to make decisions and control our impulses. Thus, adolescence is the period of development of the frontal lobes and it is when we gain control over our behavior. The prefrontal cortex is the last region of the human brain to achieve this maturation. The frontal (and largely the prefrontal) lobes play a key role in decision-making and in impulse control. Thus, it has been shown that the three developmental stages (childhood, adolescence, and maturity) differ significantly from each other. Our adolescents are not children or adults. They are adolescents. It is crucial that we understand what this stage of development means so that we may understand why adolescents drink alcohol and use drugs and why they do or do not make good decisions, so that we may prepare them to thrive in the adult world. This phase of life provides an open door and an opportunity that does not occur again in an individual's life. There are several reasons that alcohol is not a wise option for either adolescents or adults; we must create an environment in which our adolescents do not drink [8].

Several studies have shown that executive functioning plays a key role in the process of dependence and in the difficulties most people experience in ceasing the use of substances. Individuals tend to ignore the future consequences of their compulsive behavior with respect to a drug in favor of the immediate rewards associated with the drug's psychotropic effects [11]. Thus, a problem in neuropsychological functioning, especially of executive functioning, mediated by prefrontal brain regions may negatively influence treatment motivation and adherence to rehabilitation programs, increasing the chances of relapse. Hence, the executive deficit found in adolescent substance abusers may explain the difficulty that these individuals face in remaining abstinent and engaging in treatment.

This brain structure helps explain the peculiar behavior of the adolescent. A complex neural network is activated when we perform activities that bring pleasure; the constant pursuit of pleasant stimuli is associated with a "brain reward system". All behaviors reinforced by rewards tend to be repeated and learned. Biologically, this system aims to ensure survival through the motivation of behaviors such as eating, drinking and reproduction [3].

Structures comprising the limbic system (the ventral tegmental area and nucleus accumbens) play a crucial role in the expression of emotions and in the activity of the brain's reward system. Pleasure and modulation of reward occur through a reward "cascade" in which chains of neurons interact within the limbic system through several neurotransmitters. Extended drug consumption changes the rules of the latter. A deficiency in one or more neurotransmitters, especially dopamine, may overcome the sense of well-being, producing anxiety, malaise and a sense of craving for a substance that will alleviate the negative symptoms.

The dopaminergic system contains three pathways that are of great importance for understanding the neurobiology of chemical dependence. One of these is the mesolimbic-mesocortical pathway, which connects the ventral tegmental area (VTA) to the largest portion of the frontal cortex (higher mental functions) and limbic system (emotion). This dopaminergic pathway appears to be the pathway most highly associated with reward. Studies have demonstrated a close relationship between certain brain structures and reward. The nucleus accumbens and VTA apparently moderate the reward stimuli induced by psychoactive substances.

Much of the additive potential of drugs is associated with activation of the dopaminergic system. This activation may occur directly or indirectly. Psychoactive substances such as cocaine and amphetamine act directly on the dopaminergic system, while nicotine and opiates stimulate it indirectly. The natural causes that usually stimulate the reward system may increase its activity by as much as 100%. In the presence of psychoactive substances, however, this activity may be one thousand times greater.

Alcohol and other drugs of abuse also stimulate the dopaminergic system, thereby often generating pleasure that is considerably more intense than that brought about by natural functions. Because these substances initially produce euphoria and well-being, addicts experience a false sense of a beneficial effect; however, the repeated and frequent use of psychoactive substances leads to a vicious cycle that affects the brain and other organs. The activity of a minute amount of a drug that reaches the brain can substantially change behavior through the normal mechanisms of neurotransmission. A better understanding of the individual and social psychological mechanisms that underlie drug dependence is necessary. To conclude, everyone has brain systems, neurotransmitters and receptors; however, only a few individuals are susceptible to drug use and abuse [3].

According to the Brazilian Center for Information on Psychotropic Drugs (Centro Brasileiro de Informações sobre Drogas Psicotrópicas—Cebrid) of the Federal University of São Paulo (Universidade Federal de São Paulo), there has been a considerable increase in substance consumption among youth, mainly associated with the consumption of inhalants such as marijuana, cocaine and crack. However, alcohol and tobacco remain by far the drugs that are most used throughout life and with which more problems, for example traffic accidents and violence [12], are associated based on evidence that they decrease awareness and affect critical judgment, which is already altered in the adolescent by emotional processes.

The earlier drug consumption is initiated, the greater the risk of dependence, of associated mental disorders and of behavioral changes as a result of drug use. This is due to neural plasticity, which, when stimulated, results in synaptic rearrangement. Changes in neurotransmitter reuptake, particularly the reuptake of dopamine, also occur. Every drug of abuse causes the release of dopamine in the mesocortical and mesolimbic areas, thereby affecting the limbic system. Both adolescence and drug use lead to a reduction in the number of dopaminergic receptors in the reward system; when combined and allied with stress, they cause a profound reduction in the ability to activate this system. When this combination is present during a large part of an individual's adolescence, it will be consolidated in adulthood, resulting in the presence of fewer dopamine receptors. Individuals with this condition become especially prone to addiction, and their chances of developing problems related to drug abuse are greatly increased [3].

Cognitive Changes in Adolescents and Adults Suffering from Substance Dependence

Adolescence is a period of transformations and transitions. Hormone levels are altered, causing an abrupt change in sexual impulses. Anxiety is intensified, and the altered psychological milieu leaves the individual more vulnerable to drug use and to emotional disorders.

Brain transformations that occur during adolescence bring reflexive thoughts on the behavior of adolescents. Unlike heredity, age, or skeletal changes, cognition carries considerable potential for plasticity and flexibility, even in complex clinical situations that may change the patterns of brain activity [13].

Cognition is expressed through a set of complex brain processes that are responsible for building the knowledge that humans acquire and need throughout their development. Thus, mental processes are the cornerstone of perception, attention, motivation, action, planning and thinking, in addition to learning itself and memory [14].

Cognitive function is the expression of a set of complex brain processes that are responsible for building the knowledge that humans acquire and need throughout their development. Thus, when we refer to cognitive processes, we are talking about brain function, which encompasses attention, memory, language, executive functions, visual-spatial and praxis functions and perception.

The chronic use of psychoactive substances, especially when it occurs in association with other psychiatric disorders, exacerbates the deficit. Moreover, this deficit may last longer in patients who begin to use such substances early in life, i.e., during the period in which the brain is still developing [15].

Plasticity stems from processes of controlled neural maturation, i.e., typical neurobiological changes that occur beginning with the first synapses formed during the prenatal period and extending in an orderly manner until the brain's synaptic net-

Table 10.1 Brain development (Adapted from Kluger, J., ed. Your brain—A user's guide. New York: Time Books, 2009)

From conception to birth	The development of the brain and nervous system begins three weeks after conception. A refined mechanism of induction, proliferation, differentiation and cell migration enable the fetus to interact with his body and with the outside world. Between the 20th and the 30th week, the fetus is able to respond to auditory stimuli from the intrauterine environment as well as to external sounds
Childhood: from 0 to 5 years	Excess neurons and synapses are "pruned" during the first 18 months, However, the brain continues to grow, reaching 90 % of the size it will have in adulthood. Brain cells become more able to communicate, and the baby soon develops his communication skills. By the age of 3, the basic brain structure of the child is completed
Childhood: from 5 to 10 years	Significant growth of the parietal and temporal lobes occurs at this stage. Because these are the crucial brain regions for language and understanding spatial relationships, this period is excellent for learning new languages and music. Until the age of 7, the axons of the reticular formation in the brainstem are covered by a myelin sheath, vital to effective cellular communication
Puberty: from 10 to 13 years	Shortly before puberty, the volume of gray matter in the brain reaches its peak, especially in the frontal lobe, the central station of planning, impulse control and reasoning. This growth may be triggered by waves of sex hormones
Adolescence: from 13 to 20 years	The brain begins to shrink at this point as unnecessary neural pathways are "pruned". It loses approximately 20 % of its weight and volume each decade from this phase on. In adolescence, the parietal and temporal lobes, which are associated with spatial, sensory, auditory and language areas, are developed. With this, the brain will be well equipped to address social and intellectual challenges
Adulthood: from 20 to 30 years	The prefrontal cortex, a brain structure crucial for thinking and for evaluating the consequences of actions, is fully developed by the age of 30, enabling improved executive function. At this age, information processing also begins to slow down, which means that the nerve impulses are transmitted more slowly than before.
Middle age: from 30 to 60 years	Learning, memory, planning and other complex mental processes become more difficult, and reactions to stimuli consume more time. The brain is able to combat the effects of aging, offsetting declining functions to strengthen the processes of thought and memory. With this, mental activity may even improve
Old age: above 60 years	Recent studies on brain aging show that the rate of decline in brain function may be decreased by factors linked to lifestyle, such as regular exercise, healthy eating and intellectual activity, among others. One of the most important challenges for societies is related to diseases associated with aging, such as Alzheimer-type dementia

work achieves the complexity of that of an adult brain [16]. It is possible to observe this process by following brain evolution from conception to old age (Table 10.1).

Freud, cited by Aberastury and Knobel [17], highlights the difficulty of defining the boundary between normal and pathological behavior during adolescence, stating that the commotion that occurs during this period of life should be considered normal. He also emphasizes that a stable psychological equilibrium during adolescence would be abnormal because adolescence is a phase of imbalances and instabilities.

During development, changes occur in the brain and in behavior. According to Kolb and Whishaw [18], neurons are connected in an increasingly complex manner as the brain develops. These connections provide the foundation for increasing behavioral complexity.

Is it nevertheless possible to explain adolescent behavior on the basis of brain functioning? According to Herculano-Houzel [19], the adolescent brain is fundamentally different from both the infant brain and the adult brain, and the presence of differences in several brain regions may explain the behavioral changes typical of the adolescent. It all begins in the hypothalamus. According to this author [19], the sex hormones that are produced in adolescence do not encounter a mature brain that is prompt to respond. Thus, to transform an infant into an adolescent, it is not sufficient to flood the infant brain with hormones: it is necessary that the brain changes during adolescence and then respond to hormones.

The adolescent brain undergoes a chemical reorganization along with changes in the amounts of specific neurotransmitters that carry messages from one neuron to another. The different abilities of the child, the adolescent and the adult are not associated with an increase in the number of neurons; instead, the adolescent brain undergoes a chemical and structural reorganization that results in changes in its ability to exchange signals among neurons.

When facing chemical dependence, executive functions must be emphasized. The definition of executive functions includes the ability to initiate actions, to plan and predict means to solve problems, to anticipate consequences and to change strategies flexibly, monitoring behavior step by step and comparing the partial results with the original plan [6].

Upon deficit or changes in the brain's executive functions, damage is made evident through impulsive, unsystematic and unplanned behavior. There is no flexibility in thinking. Perception of reality can be intermittent, and there is no global perception. Moreover, restrictions on hypothetical-inferential thinking, difficulty in distinguishing relevant and irrelevant facts, a shortage of verbal tools by means of which to communicate appropriate responses, apathy and forgetfulness may occur. Hence, we are presented with a rather disorganized behavioral picture [6].

Executive functions allow humans to independently and autonomously develop activities directed at specific goals that are closely associated with human behavior. These functions include complex actions that depend on the integrity of several cognitive, emotional, motivational and volitional processes, all of which are intimately associated with frontal lobe functioning [17].

Thus, executive functions may be divided into four basic components:

- volition;
- planning;
- purposeful action; and
- effective performance.

Volition, which is the ability to generate intentional behavior, requires motivation, initiative and self-consciousness. Loss of volitional ability leads to significant functional impairment in which the individual may become apathetic and display no initiative.

Planning requires the ability to engage in abstract, anticipatory thinking, the ability to organize a sequence of steps, impulse control, making choices, sustaining attention and preserving memory, in addition to motivation and self- consciousness.

Purposeful action requires the ability to initiate, maintain, modify and cease sequences of complex behaviors in an integrated and orderly manner as well as the flexibility to change the perceptual, cognitive and behavioral set. As for effective performance, it comprises self-monitoring and self-regulation.

Thus, executive functioning may be defined as the ability to extract information from several verbal and nonverbal brain systems and to act on this information to produce new responses, providing guidance to the functional systems for efficient information processing [1].

Neuropsychological evaluation is the systematic analysis of cognitive, sensory, motor, emotional and social changes resulting from brain damage; it is typically performed through neuropsychological and scale tests and clinical observation [17]. Neuropsychology, which originated from neurology and psychology, is currently a field that is intermediate between neuroscience and behavioral science. Neuropsychology aims to apply evaluation and intervention principles based on the scientific study of human behavior throughout the life cycle and to determine how human behavior is associated with normal and altered CNS functioning [18]. Because substance abuse is extremely harmful to the physical, cognitive, behavioral and emotional balance of the individual, neuropsychological evaluation contributes to the recognition of possible cognitive impairments in such individuals, allowing for the identification of altered behaviors and for the development of strategies to minimize the deficits resulting from substance abuse [18]. To determine whether substance use has changed the cognitive performance of the individual or whether the damage found was present from childhood, an assessment of the individual's cognitive potential prior to use becomes necessary, i.e., pre-morbid functioning must be estimated [19]. For this, it is necessary to consider the historical process of identity construction, including past and present periods of substance use, personality characteristics, family dynamics and other cultural issues associated with substance use. Substance use and the presence of cognitive impairments may potentially produce behavioral, emotional and personality changes in individuals. The latter aspects should be considered during the process of treatment and rehabilitation [1].

Overall, neuropsychological evaluation precedes clinical intervention, allowing reflection on the cognitive and behavioral conditions of patients and an assessment of their abilities and limitations as well as the impact of sequelae in their daily activities [2]. Along with other fields of neuroscience, such as neuroimaging, neurology and psychiatry, neuropsychology involves the study of cognition, emotion, personality and behavior and focuses on the relationship between these aspects and brain functioning [20].

Evidence for Cognitive Changes Among Chronic Users of Alcohol and Other Drugs

Dependence is a disorder of brain function caused by the consumption of psychoactive substances. The use of such substances affects normal brain processes of sensory perception, emotions and motivation [21].

Neuroscience has identified neural circuits that are involved in the abuse of psychoactive substances and has identified brain regions, neuroreceptors, neurotransmitters and common neurological pathways that are affected by drugs [21].

Psychotropic substances may be divided into the following categories [21]:

- CNS depressants (e.g., alcohol, sedative-hypnotics, volatile solvents);
- CNS stimulants (e.g., nicotine, cocaine, amphetamines); and
- CNS-disturbing substances (e.g., *cannabis*, LSD).

The brain is organized into several distinct regions with specialized functions. One of these regions is the brainstem, which contains vital structures such as centers for the control of breathing and surveillance. The mesencephalon (midbrain) is a region that contains several areas that are important in psychoactive substance dependence; these regions are involved in motivation and learning processes associated with important environmental stimuli and in the reinforcement of behaviors that bring pleasant and essential consequences, such as eating and drinking. The prosencephalon (forebrain) is more complex, allowing abstract thinking, planning, thoughts, associations and memories [21].

It has been shown with imaging techniques that specific forebrain regions are activated by stimuli that induce in the dependent person an urgent need to consume a particular substance and that other regions in the brains of such individuals function abnormally after acute or chronic ingestion of psychotropic substances and/or under conditions of established dependence. Psychotropic substances may mimic the effects of natural or endogenous neurotransmitters or interfere with normal brain function by blocking neurotransmitter function or by altering the normal processes of accumulation, release and removal of neurotransmitters. An important mechanism of action of these substances is blockage of the reuptake of a neurotransmitter after its release at the presynaptic terminal. Reuptake by the presynaptic membrane is a normal mechanism of neurotransmitter removal at the synapse. When such reuptake is blocked, the normal effects of the neurotransmitter are exacerbated [21].

Psychoactive substances that bind to and strengthen the functions of receptors are called agonists, whereas those that bind to receptors in such a way as to block the normal function of the receptor are called antagonists [21].

Substance consumption and drug dependence represent an important public health problem. The 2002 WHO report on world health indicated that 8.9% of the global burden of diseases results from the consumption of psychoactive substances. Tobacco, alcohol and illicit drugs represent, respectively, 4.1, 4 and 0.8% of this burden [21].

According to studies performed by the Einstein Center for Alcohol and Drugs at the Israeli Hospital Albert Einstein (Núcleo Einstein de Álcool e Drogas-NEAD do

Hospital Israelita Albert Einstein), the main harmful effects of substance consumption can be divided into four categories [21]:

- chronic effects—liver cirrhosis, lung cancer and emphysema, as well as damage caused by sharing needles (Human Immunodeficiency Virus (HIV) and hepatitis B and C viruses);
- acute or short-term effects—overdose, traffic accidents, suicide and aggression;
- harmful effects—serious social problems such as separation or arrests; and
- chronic social problems, such as losses associated with work or family life.

In other studies, crack dependents presented great loss in memory functions, whereas alcohol dependents displayed impairments in attention, memory, learning and overall executive function, and individuals dependent on snorted cocaine showed decreased verbal and phonological fluency [22]. Compared with nonusers, users of alcohol, cocaine and crack present greater impulsivity, impaired sustained attention and decreased verbal retention in tasks that require time and preparation or that involve difficult reasoning and abstraction [23].

Different psychoactive substances act differently to produce their effects in the brain. They bind to different receptor types and may increase or decrease the activity of neurons through several mechanisms. As a consequence, they have distinct effects on behavior and distinct rates of tolerance development, abstinence symptoms and short- and long-term effects.

Cognitive deficits are common in individuals whose brains are exposed to psychotropic substances. A few cognitive effects may be considered acute, i.e., they are present for a short period, whereas other effects may be considered chronic because they may persist for weeks, months or years after the last exposure to the drug and may or may not be reversible [24].

The main neuropsychological findings associated with the use of licit and illicit substances are briefly described below [21].

Alcohol

Among the global losses associated with alcohol abuse and dependence are

- violent deaths;
- exposure to risk behaviors;
- cognitive-behavioral deficits; and
- emotional deficits and violence.

As the concentration of alcohol in the blood increases, the effects of intoxication become more intense and may affect cerebellar function, causing imbalance and difficulties associated with coordination and speech articulation. Higher doses of alcohol may cause loss of consciousness; if blood levels reach 0.5 %, there is risk of death from respiratory depression [25]. Chronic alcohol use can lead to brain damage, such as that observed in Wernicke-Korsakoff syndrome, a disorder characterized by significant loss of memory, motor and sensory deficits, dementia, profound amnesia for recent and past events, disorientation in time and space and lack of insight. This disorder is caused by a deficiency in thiamine (vitamin B1); a few studies have reported associated subcortical lesions and cortical atrophy [26]. Cognitive impairment associated with alcohol use has been consistently associated with memory, visual-spatial organization, psychomotor problems and decision-making function. Furthermore, severity of alcohol use has been consistently associated with decreased performance on tests that evaluate executive functioning and with damage to distinct regions of the prefrontal cortex [27].

Tobacco

Tobacco use is increasing rapidly in developing countries and among women. Currently, 50 % of men and 9 % of women in developing countries smoke compared to 35 % of men and 22 % of women in developed countries [21].

The effects of long-term tobacco use on health are well known. Noteworthy is the difficulty of dissociating the effects of nicotine from the effects of other components of tobacco [21].

Illicit Psychoactive Substances

Data from the United Nations Office on Drugs and Crime (UNODC) show that large-scale dealing of cocaine, heroin, *cannabis* and amphetamine-type stimulants occurs in several parts of the world. The availability of cocaine, heroin and *cannabis* depends on the level of cultivation in producer countries and on the success or failure of trafficking organizations. However, even with increasingly efficient law enforcement, users apparently continue to have access to sufficient supplies of these substances. According to UNODC reports, approximately 200 million people consume at least one type of illicit substance [28].

Marijuana

Depending on the dose of smoked or orally ingested marijuana in acute use, there are changes in psychomotor speed, perceptual changes, difficulties with immediate memory and in data recovery after a short time and, especially, decreased performance on tasks that involve complex cognitive performance, indicated by changes in attentional focus, inhibition of automatic responses and decreased ability to abstract [21].

Cocaine

Cocaine is a stimulant. Cocaine use causes acceleration in thinking speed, psychomotor restlessness (from difficulty in remaining still to more serious episodes of restlessness), increased alertness and appetite inhibition. Mood changes are subject to great variability, ranging from euphoria (disinhibition, loose talk) to symptoms of psychological discomfort (fear, anxiety and speech inhibition) [21].

Cognitive impairment, abnormalities in specific regions of the cortex, failure in motor function and decreased reaction time have been found[21].

Amphetamines

Amphetamines stimulate CNS activity, i.e., they cause the brain to work faster, leaving people more "lit", “connected”, and able to function on less sleep, among other effects.

The effects of amphetamines are broad, influencing several behaviors. A person under the influence of amphetamines typically experiences insomnia and loss of appetite, feels full of energy and talks rapidly, seeming to be “electric”.

Prolonged use of amphetamines causes sleep disturbances, anxiety, loss of appetite, changes in brain dopamine receptors, localized metabolic abnormalities and motor and cognitive shortcomings [21].

Ecstasy

In addition to its hallucinogenic effect, which is characterized by changes in time perception, a decreased sensation of fear, panic attacks, psychosis and visual hallucinations, ecstasy has stimulant effects that include increased heart rate and blood pressure, dry mouth, nausea, sweating and euphoria. The active ingredient in ecstasy, 3,4-methylenedioxy-methamphetamine (MDMA), is a drug that in addition to causing hallucinations may also lead to a state of excitement, which is doubly dangerous.

Prolonged consumption of ecstasy results in damage to serotonergic brain systems and to behavioral and physiological complications as well as to physical and psychiatric problems, such as impaired memory, impaired decision-making and self-control, paranoia, depression and panic attacks [21].

Hallucinogens

Hallucinogens induce hallucinations and delusions. It is important to note that these effects vary considerably, i.e., they depend on conditions such as the sensitivity and personality of the individual, the expectations the person has about the effects, the environment and the presence of other persons, as well as on other factors.

Sometimes the effects of hallucinogen use are pleasant (“good trip”); in such cases, the person feels rewarded by experiencing unusual sounds, bright colors and hallucinations. On other occasions, the mental phenomena associated with hallucinogens are unpleasant and may include terrifying visions, sensations of bodily deformation or a certainty of imminent death, among other effects; these are the "bad trips". Both "good" and "bad" trips may be guided by the environment and by the previously cited factors.

Acute or chronic psychotic episodes and flashbacks or relapses of the substance's effects long after consumption are a few of the effects of prolonged hallucinogenic drug consumption.

Rehabilitation of Preserved Brain Functions

The results of a neuropsychological evaluation not only indicate the existing damage but also reveal the functions that are preserved after the prolonged consumption of substances [29]. This latter aspect is of great relevance because the preserved functions provide support for the rehabilitation of impaired functions that, as they are stimulated, positively interfere with the individual's adherence to drug treatment.

Patients enrolled in cognitive rehabilitation programs show greater commitment to treatment and higher percentage of abstinence days after treatment than patients who do not participate in such programs. Considering that cognitive deficit among patients with substance abuse has important implications, the authors note the need to engage these patients in a variety of therapeutic approaches that include contact, coding and the incorporation of novel information that will help them initiate and execute plans to reorganize their behavior throughout treatment. Thus, an improvement in the cognitive process directly involves behavioral changes [29].

Neuropsychological rehabilitation is broadly based because, in addition to the rehabilitation of cognitive aspects of mind, it addresses emotional issues and behavioral changes, integrating psychotherapy with counseling of the patient's family members [6].

Thus, when addressing issues relating to adolescents, one penetrates an extremely wide field rich in important specific aspects of the adolescents' life histories, including both intrinsic issues (biological and psychological factors) and extrinsic issues (family social and emotional relationships).

In this chapter, we address a small part of the immense universe that surrounds adolescents, especially those that excessively use alcohol, tobacco or other drugs [30].

References

1. Diehl A, et al. Dependência química: prevenção, tratamento e políticas públicas. Porto Alegre: Artmed; 2011.
2. Johnston LD, O'Malley PM, Bachman JG, et al. Monitoring the future national survey results on drug use, 1975–2006: vol. I, secondary school students. Bethesda: National Institute on Drug Abuse; 2007.
3. Soares HLR, Gonçalves HCB, Werner Jr J. Cérebro e o uso de drogas na infância e adolescência. Fractal Rev Psicol. 2010;22(3):239–240.
4. Bechara A, Dolan S, Denburg N, et al. eds. Decision-making deficits, linked to a dysfunctional ventromedial prefrontal cortex, revealed in alcohol and stimulant abusers. Neuropsychologia. 2001;39:376–89.

5. Goldstein RZ, Volkow ND. Drug addiction and its underlying neurobiological basis: neuroimaging evidence for the frontal cortex. Am J Psychiatry. 2002;159:1642–52.
6. Cunha PJ, Novaes M. Avaliação neurocognitiva no abuso e dependência do álcool: implicações para o tratamento. Rev Bras Psiquiatr. 2004;26(Supl I):23–7.
7. Verdejo-Garcia A, Lopez-Torrecillas F, Arcos FA, et al. Differential effects of MDMA, cocaine, and cannabis use severity on distinctive components of the executive functions in polysubstance users: a multiple regression analysis. Add Behav. 2005;30:89–101.
8. White AM. Alcohol: una amenaza para el desarrollo neuropsicológicode los adolescente. Rev Farmac. 2007;330:11–3.
9. American Academy of Pediatrics: The classification of child and adolescent mental diagnosis in primary care: diagnostic and statistical manual for primary care (DSM-PC) child and adolescent version. Elk Grove Village: American Academy of Pediatrics; 1996.
10. Barbizet J, Duizabo PH. Manual de neuropsicologia. Porto Alegre: Artes Médicas; São Paulo: Masson; 1985.
11. Volkow ND, et al. Dependentes de cocaína apresentaram uma resposta atenuada à intoxicação alcoólica em regiões do cérebro límbico. Vida Sci. 2000;66:PL161–7.
12. Galduróz JCF, Noto AR, Carlini EA. Tendências do uso de drogas no Brasil: síntese dos resultados obtidos sobre o uso de drogas entre estudantes do 1° e 2° graus em dez capitais brasileiras. São Paulo: Cebrid; 1997.
13. Fonseca V. Cognição, neuropsicologia e aprendizagem: abordagem neuropsicológica e psicopedagógica. 4. ed. Rio de Janeiro: Vozes; 2009.
14. Vieira EB. Manual de gerontologia: um guia teórico-prático para profissionais, cuidadores e familiares. Rio de Janeiro: Revinter; 1996.
15. Jager G, Ramsey NG. Long-term consequences of adolescence cannabis exposure on the development of cognition, brain structure and function: an overview of animal and human research. Curr Drug Abuse Rev. 2008;1:114–23.
16. Gage FH. Structural plasticity of the adult brain. Dialog Clin Neurosci. 2004;6:135–42.
17. Aberastury A, Knobel M, orgs. Adolescência normal: um enfoque psicanalítico. Porto Alegre: Artes Médicas; 1981.
18. Kolb B, Whiswaw I. Neurociência do comportamento. Barueri: Manole; 2002.
19. Herculano-Houzel S. O cérebro em transformação. Rio de Janeiro: Objetiva; 2005.
20. Kolling NM, Silva CR, Carvalho JCN, et al. Avaliação neuropsicológica em alcoolistas e dependentes de cocaína. Aval Psicol. 2007;6(2):127–37.
21. Hospital Israelita Albert Einstein. Núcleo Einstein de Álcool e Drogas (NEAD). Álcool e drogas sem distorção. Disponível em: www.einstein.br/alcooledrogas/.
22. Oliveira MS, Laranjeira R, Jaeger A. Estudos do prejuízo cognitivos na dependência do álcool. Psicol Saúde Doença. 2002;3:205–12.
23. Nassif SLDS. Aspectos neuropsicológicos associados ao uso de cocaína. In: Andrade VM, Santos FHD, Bueno OFA. Neuropsicologia hoje. São Paulo: Artes Médicas; 2004. p. 371–83.
24. Peuker AC, Fogaça J, Bizarro L. Expectativas e beber problemático entre estudantes universitários. Psicol Teoria Pesq. 2006;22(2):193–200.
25. Pliszka SR. Neurociência para o clínico de saúde mental. Porto Alegre: Artmed; 2004.
26. Edward G, Marshall EJ, Cook CCH. O tratamento do alcoolismo: um guia para profissionais da saúde. 3rd ed. Porto Alegre: Artmed; 1999.
27. Verdejo-Garcia A, Lopez-Torrecillas F, Gimenez CO, et al. Clinical implications and methodological challenges in the study of the neuropsychological correlates of cannabis, stimulant, and opioid abuse. Neuropsychol Rev. 2004;14(1):1–41.
28. National Institute on Drug Abuse (NIDA). Principles of drug addiction treatment and research-based guide. 2nd ed. Bethesda: National Institute on Drug Abuse; 1999.
29. Banks ME. The role of neuropsychological testing and evaluation: when to reefer. Adolesc Med. 2002;13:643–62.
30. Figlie NB, Bordin S, Laranjeira R. Aconselhamento em dependência química. 2nd ed. São Paulo: Roca; 2010.

Rafaela Larsen Ribeiro Biologist. Specialization and MS in Pharmacology. Doctor of Science, Department of Psychobiology, Federal University of São Paulo (UNIFESP). Researcher and Co-advisor, Department of Psychobiology, UNIFESP.

Shirley Yamaguchi Clinical psychologist specializing in Cognitive-Behavioral Therapy. Specialist in Chemical Dependency. Specialist in Neuropsychology. Participant in Course in Cognitive Behavioral Therapy, Institute of Psychiatry, School of Medicine, University of São Paulo (USP).

Silvana Pochetto Cracasso Educator. Educational psychologist. Specialist in Neuropsychology. Graduate student in Chemical Dependency.

Fernanda Peralta Educator specializing in Chemical Dependency, School of Medicine, University of São Paulo—USP. Teacher of Projects on Sexuality and Prevention of the Abuse of Alcohol and Other Drugs in private schools. Graduate studies in Clinical and Educational Psychopedagogy.

Elton Rezende MD, Federal University of Minas Gerais (*Universidade Federal de Minas Gerais*—UFMG). Specialist in Psychiatry, Brazilian Psychiatric Association. Specialist in Chemical Dependency, Federal University of São Paulo (UNIFESP). Graduate studies in Cognitive-Behavioral Therapy, University of São Paulo (USP).

Chapter 11
Relationships Between Mood Disorders and Substance Abuse During Adolescence

Luís Pereira Justo and Helena Maria Calil

Introduction

Mood is the overall affective quality that is based on how the individual experiences the world and themselves in a given time or period. It is the affective tone of this experience. Typically, the states of mood fluctuate according to internal or external, objective or subjective events within a certain range compatible with such events, and these oscillations are inherent to events. Therefore, mood swings can be considered normal. However, they can also occur for pathological reasons and represent a significant change in the functioning of the affected person. When mood swings are pathological, they lose the quantitative and qualitative proportionality in relation to the events that would have promoted them and are no longer explained by them. Furthermore, these oscillations can arise spontaneously, without association with any event; they can be durable and disruptive and can destabilize the individuals and their relationship with the world. Mood disorders can occur in several mental disorders but are more common in depressive and bipolar disorders.

This chapter will focus on mood disorders (MDs) that occur during adolescence and their relationship to substance abuse. These substances refer to licit substances, such as alcohol and tobacco, and to illicit substances, which include marijuana, cocaine, crack cocaine, oxi, several opioids, amphetamines and their derivatives, such as ecstasy and methamphetamine, lysergic acid (LSD), inhalants, and ketamine, among others.

L. P. Justo (✉)
Associação Fundo de Incentivo à Pesquisa, São Paulo, Brazil
e-mail: lpjusto@gmail.com

H. M. Calil
Department of Psychobiology, UNIFESP/EPM, São Paulo, Brazil
e-mail: hmcalil@unifesp.br

D. De Micheli et al. (eds.), *Drug Abuse in Adolescence*,
DOI 10.1007/978-3-319-17795-3_11

Mood Disorders

MDs represent complex alterations of the normal functioning of individuals. When considered together, they are most likely heterogeneous conditions in both clinical manifestations and the biologic determinants, which have the pathological modifications of the affective quality of experiences as common denominator.

MDs can occur episodically, with periods of acute symptoms and normal or relatively normal periods. However, it is not unusual for the symptoms to persist between acute episodes, which will then prevent the resumption of normal functioning of the individual, eventually causing significant disability.

The diagnosis of the classical conditions is usually relatively easy, but both unipolar depressions and the episodes of bipolar disorder may have different presentations and little evidence when only the most paradigmatic characteristics are considered. Therefore, identifying these disorders can be challenging.

Guidelines and diagnostic criteria are available in the two major classifications of mental disorders, the International Classification of Diseases, Tenth Revision (ICD-10) and the Diagnostic and Statistical Manual of Mental Disorders, Fourth Edition (DSM-IV-TR); however, these classifications provide features that are quite limited in clinical practice. The diversity of symptomatic manifestations and contexts in which they occur makes the evaluation process more complex and nuanced. Depression is often expressed indirectly as pain complaints that cannot be explained by somatic diseases, excessive food consumption behavior, or alcohol abuse, among other possibilities that frequently are not identified by the clinician as a possible underlying depressive episode. Symptoms of hypomania may also be unnoticed because they might be mistaken for variations of normal behavior when the occurrence of previous mood disorders is not properly investigated.

The severity of these diseases varies among individuals and in the same individual throughout the evolution of the disease. It should be considered that the abundance of symptoms is not always proportional to the problem they may cause regarding the impairment of the normal functioning of patients.

This chapter will use the MD descriptions according to the DSM-IV-TR for a more general and simplified view. However, as previously mentioned, the guidance provided by this classification does not cover the full complexity of these diseases, and the mentioned criteria should not be automatically used.

Major depressive disorder is characterized by the existence of at least one major depressive episode during the life (usually there are more than one episode), without the occurrence of manic, hypomanic, or mixed episodes at any time [1]. One major depressive episode is defined by the presence of five or more of the following symptoms for at least 2 weeks, in addition to sad mood or loss of interest and pleasure in activities that are normally enjoyable:

- unintentional weight loss or gain
- insomnia or hypersomnia
- agitation or psychomotor retardation
- fatigue

- loss of energy
- feelings of excessive guilt or worthlessness
- thoughts of death
- suicidal thoughts

These symptoms must represent a change in the normal functioning of the person [1].

In dysthymia, there are periods of depressed mood of at least 2 years in adults and 1 year in children or adolescents (irritable mood), with depression-free intervals lasting no longer than 2 months and depressive symptoms less severe than episodes of major depression; the absence of any manic, hypomanic, or mixed episodes are among the diagnostic criteria [1]. Symptoms may include sadness (or irritability in children and adolescents), abnormally increased or decreased appetite, insomnia or hypersomnia, low energy, fatigue, low self-esteem, poor concentration, and difficulty making decisions[1].

Depressive disorders not otherwise specified include episodes of mood swing with depressive features that do not meet the criteria for major depressive disorder, such as minor depressive disorder, post-schizophrenia depression, recurrent brief depressive disorders, and premenstrual dysphoric disorder, among others.

Bipolar disorders are divided into type I and type II. Bipolar disorder type I is characterized by at least one manic episode or one mixed episode (very frequently, but not necessarily, more than one manic episode and several major depression episodes) [1]. Bipolar disorder type II is diagnosed when there are at least one major depressive episode and one hypomanic episode, without the occurrence of manic or mixed episodes [1]. A manic episode lasts at least 1 week (or less if hospitalization is required) and is characterized by an abnormal expansive or irritable mood and three or more of the following symptoms:

- inflated self-esteem or grandiosity
- a decreased need for sleep
- pressured speech
- racing thoughts, which in severe cases lead to flights of ideas
- poor concentration due to distractibility
- an increase in physical activity and task execution (usually incomplete and deficient)
- psychomotor agitation
- an increased libido that can decrease the critical capacity and lead to embarrassing and risky situations of exposure;
- excessive involvement in pleasurable activities
- purchases or investments that are often excessive, unnecessary, or inconsistent with the finances of the person

It is important to note that these symptoms are not part of the usual personality of the individual [1]. In the mixed episodes, there are symptoms of both major depression and mania, and they last at least 1 week.

In cyclothymic disorder, people can have several hypomanic symptoms and several depressive symptoms that last at least 1 years (1 year for children and adolescents), with symptom-free intervals lasting no longer than 2 months; they have insufficient symptoms to meet the criteria for major depressive episode disorder or manic or mixed episodes [1].

Several diseases can occur concomitantly with mood symptoms and eventually cause confusion regarding the origin and nature of these symptoms. Some of these diseases are stroke, Parkinson's disease, hyper- or hypothyroidism, hepatitis, several infectious diseases, and vitamin B12 deficiency, among others. Ingested substances, such as medications, food, and illicit drugs, can also cause significant changes in mood.

MDs usually occur in traumatic situations that are difficult to manage for the patients and for the people who live with them. It may be very difficult for the individuals more directly involved to understand and accept the behavior changes caused by these syndromes. Usually, people presenting recurrent episodes of depression, mania, or hypomania lose their capabilities for social interactions and job performance, even during the intercritical periods. These diseases can lead to greater difficulties when they occur during childhood and adolescence compared to their occurrence during adulthood. These difficulties include the associations of a child or adolescent with their identification and treatments and their impact on the acquisition of adaptive skills necessary for life [2].

The recognition of the depressions as illnesses comparable to other conditions that impair the health of individuals was formally recorded a few centuries earlier for adults than for children and adolescents, which demonstrates the difficulties that still remain in diagnosing these disorders in young individuals [3]. The bipolar disorders have similar historical status in this population.

MDs frequently occur concomitantly with health problems that arise independently but influence each other, known as comorbidities. Comorbidities should be identified and promptly treated. Among the most important comorbidities in individuals with MDs are abuse of and dependence on substances such as alcohol, marijuana, cocaine, crack cocaine, and other drugs. While not being a morbid condition that can be considered a comorbidity, the mere use of alcohol and/or drugs in the presence of mood disorders should be considered a potential risk.

Major depressive disorders are treated with antidepressants during a limited period of time and when only a few episodes have occurred. Eventually, when the episodes become frequent and recurring, these disorders are continuously treated. For cases that do not require continuous treatment, the antidepressants should be withdrawn slowly after a few months of complete reduction of the symptoms [4]. Psychotherapy combined with pharmacological treatments can be very useful. Treatments for episodes of major depressive disorders usually last between 6 months and 1 year. Antipsychotics are required for severe depressions with psychotic symptoms until these symptoms are controlled [4]. More infrequently, drug combinations, electroconvulsive therapy, transcranial magnetic stimulation, vagus nerve stimulation, or deep brain stimulation, among other techniques, may be necessary in cases of treatment-refractory depression [5].

The basic pharmacological treatment for bipolar disorders is mood stabilizers, which include very different substances that have demonstrated this property. Lithium is the main mood stabilizer, but sodium valproate or valproic acid, lamotrigine, and carbamazepine are also widely used [6]. Recently, certain atypical antipsychotics have also been used as mood stabilizers [6]. In addition to the drugs that act as stabilizers and are commonly used in long-term treatment, antidepressants, antipsychotics, and benzodiazepines can be used, particularly as pharmacological adjuvants during the episodes of depression or mania. Psychosocial interventions combined with the pharmacological treatment may also be very useful.

Mood Disorders During Adolescence

Diagnosis

Adolescence is a period in which the human being is naturally subjected to an overall instability of personality and behavior due to the major biological, psychological, and social changes required for the adaptive demands of the transition from childhood to adulthood. Mood syndromes similar to the ones affecting adults can occur during this period of life and constitute episodes of depression, mania, or hypomania. However, the presentations of pathological mood changes during adolescence may have particularities that usually complicate their diagnosis [7]. Additionally, it has been demonstrated that the consequences of mood disorders are more severe when they occur before adulthood, with higher rates of chronicity, more comorbidities, and higher risk exposure [8].

Mood syndromes in this stage of development, as well as in adulthood, may occur in several nosological contexts, both primary and secondary. In general, mood syndromes can be secondary to different types of medical conditions as well as to the use of substances, either medications or recreational drugs (including alcohol and other drugs). In addition, mood syndromes can be primary manifestations of major depressive disorder and other manifestations of depressive syndromes (dysthymia), bipolar disorder (including the bipolar spectrum disorders), or cyclothymia [9, 10].

Recently, the number of children and adolescents diagnosed with bipolar disorder has increased significantly compared to previous decades. This fact might be due to the greater attention devoted to the mood symptoms in this population [11], or alternatively, it might be a real increase in the prevalence of this disorder among the young population [7]. The diagnosis of adolescent depression can be highly complex and is not limited to examining the lists of symptoms but rather relies on the use of broad information about the familial and environmental context, with an eventual need for multidisciplinary assessments that address the patient from different perspectives [12, 13]. ICD-10 and DSM-IV-TR may be insufficient, but they should not be ignored. The application of the ICD-10 guidelines and DSM-IV-TR criteria seems to be indicated, except for the depressed mood item that is less easy to characterize in these conditions [14]. To demonstrate that studies may produce different results and conclusions may be unclear when analyzed together, one study demonstrated that depressed mood and sadness (to a slightly smaller degree) are the most frequent symptoms in major depression during adolescence [15].

Episodes of major depression, bipolar depression, mania, hypomania, and mixed in adolescents may present different characteristics compared to episodes observed in adults. For instance, irritability in this age seems to be a manifestation very common in both depressions and mania and hypomania, and due to its predominance in these syndromes, it seems to be the only major change compared to the usual behavior. For adults, this symptom is usually not sufficient to lead to the diagnosis but is rather a component of a set of typical symptoms of the disorder [1]. Irritability in an adolescent with mania presents more aggressive features, with more frequent "attack" behaviors than observed in adults [16]. Still comparing with adults, it also appears that adolescents exhibit a higher exacerbation of sexuality and impulsiveness during the manic episodes.

It is possible that bipolar disorders develop more gradually in adolescents than in adults, which often lead parents to believe that the different behaviors are due to the natural changes in adolescence [17].

Prevalence

Epidemiological studies clarifying the prevalence of MDs during adolescence, especially in developing countries, are scarce. However, descriptive information from developed countries reveal the prevalence of these disorders worldwide [3]. Prevalence rates of major depressive disorder among adolescents vary from 0.7 to 9.8% in the early stages and 15–25% in late adolescence [3]. Regarding bipolar disorder, studies show wide variation in prevalence rates; the prevalence of the subthreshold symptoms for the diagnosis ranged from 6 to 13.3%, while the prevalence of the full bipolar disorder symptoms ranged from 0 to 1% [11].

Comorbidities

The presence of comorbidity in adults with MDs is relatively common, and it is likely that the same is true for adolescents. Studies report the presence of sufficient symptoms for the diagnosis of attention deficit/hyperactivity disorder (ADHD) in 90% of adolescents diagnosed with bipolar disorder; furthermore, the diagnoses of different anxiety disorders, especially obsessive-compulsive disorder, disruptive behavior disorder, and substance use, are described as common comorbidities [18]. In addition, panic disorder seems to be common among adolescents with bipolar disorder and exhibits a significantly negative impact on the evolution of the disorder [19]. There may be some confusion between the symptoms of bipolar disorder and other pathological conditions due to their similarities, which can make it difficult to know whether there is only one or more illnesses occurring simultaneously. Some common comorbidities that occur in adolescence may mask the diagnosis of a primary mood disorder, such as the simultaneous occurrence of ADHD, oppositional defiant disorder or conduct disorder as well as substance abuse [16, 20]. The frequency of the occurrence of symptoms, which is more common in bipolar disorders than in recurrent major depressive disorder, should be analyzed when there are doubts about the comorbidities or the diagnosis of primary MDs. Other aspects that may be relevant for differential diagnosis or the detection of comorbidities are the presence of clearly identifiable cases of MDs in the family and monitoring for a sufficient period of time to identify the symptoms, as a cross-sectional evaluation might be not able to identify the disorder due to the great similarity in the symptoms of the mentioned diagnosis and the manic or mixed episodes in bipolar I disorder [21].

Impact

The speed and accuracy of diagnosis are essential for providing proper treatment and minimizing damage because the impact of MDs on the development of very young people can be dramatic. It has been shown that the morbidity is higher for individuals who have bipolar disorder since childhood and adolescence compared to individuals who became ill as adults. This fact emphasizes the importance of early detection for introducing proper interventions [22]. One study showed that suicidal thoughts might reach 76 % and suicide attempts 31 % among children and adolescents with bipolar disorder [7].

Treatment

Adolescents usually do not face life as adults do, and their expectations of immediacy are characteristic of this stage, which in general means seeking satisfaction and accepting limits. Therefore, it is expected that the treatment of a disease in this stage of life can be more difficult than in adults. In addition, these diseases that impair important capabilities for proper mental function may be more harmful in this population, which is still in development.

Multimodal therapeutic approaches are desirable. Pharmacological treatment is required, but its combination with psychosocial interventions, such as psychotherapy, psychoeducational techniques, and social integration activities, can also be very useful. The inclusion of the family in the therapeutic processes is also important because the adolescents at this age often live with their nuclear families, and the inter-relationship is usually more intense than among adults.

Basically, adolescents use the same drugs used by adults, including mood stabilizers, such as lithium, sodium valproate/valproic acid, carbamazepine, and lamotrigine. In addition, antidepressants and typical or atypical antipsychotics may be helpful. However, the doses need to be adjusted according to the age and developmental conditions of the patients, and care for the adverse events should be even stricter than in adults.

Relationships Between Mood Disorders and Substance Abuse in Adolescents

Epidemiological studies show that the magnitude of the association between abuse of or dependence on alcohol and other drugs and MDs in adults is highly significant, especially for bipolar disorders [23, 24]. An epidemiological study in adults conducted by Conway et al. demonstrated that the diagnosis of drug dependence is more strongly associated with MDs than the diagnosis of drug abuse [25]. In adolescents, MDs are risk factors for substance abuse [26, 27]. A study conducted by Perlis et al. with 1000 adults with bipolar disorder showed that the onset of MDs during childhood and adolescence has a significantly stronger correlation with substance abuse than onset in adulthood [22].

Adolescence is characterized by behaviors that involve increased risk-taking in inter-relationship with the environment, which seems to be a natural adaptive occurrence in which individuals can experience the world by themselves and create their own differentiation in relation to their parents and caregivers [28]. Complex phenomena are involved in these adaptive processes, and several theoretical models have been proposed to explain them. The mechanisms that determine choices for the active search or avoidance of objectives and objects that develop or occur during the life and can be defined as motivated behaviors are important aspects in the individual destinies. These mechanisms seem to rely on the neurobiological basis associated with the striatum, amygdala and medial prefrontal cortex circuits [28]. The associations between pathological mood swings and motivated behavior functions can be close and biunivocal [29].

Impulsive behaviors, in which the ability to determine what is the best choice for a certain action is greatly reduced or abolished, seem to be more frequent during adolescence and are also more common among individuals with bipolar disorder. Considering that among the MDs, bipolar disorders show a high comorbidity rate with substance dependence disorders, the fact that there are changes in functions of motivation, reward, and control systems at the onset of behaviors in both pathologies should be considered [30]. Therefore, the risk of impulsive behaviors and the attraction to novelty and experimentation may be greater among adolescents with bipolar disorder compared to adolescents without MDs. This fact indicated that adolescents with bipolar disorders are more prone to substance use, including alcohol, cocaine, crack cocaine, marijuana, and several other substances[26, 27, 31].

There is evidence in the literature indicating that the substance abuse among patients with MDs may be associated with a search for self-medication [32]. Furthermore, the propensity to experience the sensations provided by mental changes caused by drugs and alcohol also seems to be important [32]. Additionally, it has been proposed that individuals with MDs are more sensitive to substance dependence disorders [30].

Basic research has suggested a relationship between the neurobiological mechanisms associated with MDs and substance abuse and dependence. This relationship is raised hypothetically in two ways: the two types of disorders could be different expressions of common neurobiological abnormalities, or substance use disorders could produce neuroadaptive changes that induce the same mechanisms associated with the occurrence of MD symptoms [33]. Data obtained from human neuroimaging studies have also assisted in the construction of hypotheses linking depressions and substance dependence because the same neurobiological alterations occur in areas involving the limbic-frontal [33]. The dopaminergic reward system might be dysregulated in individuals with major depressive disorder, which also occurs in substance abusers [34]. A reduction in the nucleus accumbens, a component of the mesolimbic-cortical reward system functionally altered by almost all MDs, was associated with the intensity of anhedonia in depressed individuals[35, 36]. Certain studies suggest some genetic overlap between bipolar disorders and substance dependence, further supporting the hypothesis that both disorders may be expressions of the same neurobiological vulnerabilities [37]. A relationship between the symptoms of mania and the actions of the stimulant substances, such as cocaine and amphetamines, has also been suggested because there are, in both cases, changes in neurobiological systems involving dopamine and noradrenaline [30]. Moreover, individuals with bipolar disorder seem to have an amplified response to amphetamine compared to individuals without this disorder [30].

Prevalence of Comorbidity

The literature has highlighted the importance of comorbidity between bipolar disorder and the problems with drug use in the adult population and its terrible consequences on the evolution of MDs, such as intensified symptoms of the episodes, higher frequency of acute episodes, persistent symptoms among episodes with higher morbidity and mortality, and lower adherence to treatment [38, 39]. Youth-onset bipolar disorder (childhood or adolescence) seems to confer an even greater risk of substance use disorders in comparison with adult-onset bipolar disorder [40]. However, there are few global studies addressing comorbidity in MDs that occur in childhood and adolescence, especially ones addressing the prevalence of substance abuse to define a clearer overall frequency of this comorbidity in young people [40].

In a study with adolescents diagnosed with some form of bipolar disorder, Goldstein et al. demonstrated that the prevalence of substance use disorders was 16% for life (sample of 249 individuals); marijuana use was the most prevalent substance, with rates of 12% among all the adolescents in the sample and 73% among individuals diagnosed with certain substance use disorders [41]. Alcohol abuse or dependence occurred in 8% of the sample, and the use of marijuana and alcohol combined occurred in 5% of the entire sample and in 30% of those diagnosed with some substance use disorders [41]. The prevalence of the use of all other substances together did not exceed 3% [41].

In a follow-up study of patients diagnosed with bipolar disorder type I in childhood or early adolescence, Geller et al. found that after 18 years of age, 35.2 % of patients presented comorbidity with disorders associated with substance use [42].

In a longitudinal study, Wittchen et al. analyzed the associations of marijuana use or disorders associated with marijuana abuse/dependence with other psychiatric disorders. This study was conducted in Germany with a population sample of adolescents aged between 14 and 17 years old, and 3 follow-up examinations were performed within 10 years. The authors found marijuana use in 67 % of the adolescents who also had symptoms of bipolar spectrum, and among those, 22 % met the criteria for abuse or dependence; the use of marijuana was 22 % among the adolescents without other psychiatric disorders, and disorders of abuse or dependence together were found in 1 % [43]. In the progress to adulthood, the risk of incidence of marijuana abuse/dependency disorder was three times higher among individuals with symptoms of the bipolar spectrum than among individuals from the control group [43].

In a review on this subject, O'Neil et al. found that the rates of comorbidity of depressive syndromes in adolescents diagnosed with substance abuse or dependence ranged from 11 to 47.9 %, with the second comorbidity being more prevalent in these conditions [44]. The same authors mention that in the opposite sense, i.e., prevalence rates of subsequent onset of substance abuse among individuals who had depressive syndromes early in life, the available evidence is not informative, nor are the data about what disorder occurs first, which could indicate whether one disorder could be a risk factor for the other [44].

It is worth mentioning that people who consume high amount of alcohol and other drugs can have mood deregulation and present symptoms that would be mistaken for a primary mood disorder. This fact might have some relevance in studies of the prevalence of comorbidity between MDs and substance abuse, resulting in misinformation about the existence of the two disorders when, in fact, there is only one [45].

Impact of Comorbidity

According to evidence obtained from clinical studies, mainly in adults, it seems that the comorbidity between MDs and abuse of or dependence on alcohol and other drugs increases the level of morbidity in both cases. Neurobiological changes resulting from the substance use that interfere harmfully in the manifestations and response to treatment of individuals with MDs might occur, and it is possible that the occurrence of MDs can increase the susceptibility to use and dependency on alcohol and several other drugs, in addition to worsening the consequences of the consumption of substances in patients with MDs compared to patients without MDs [30]. However, there is not sufficient scientific data, especially for the population under 18 years old, to make categorical statements about the correlation between these two types of psychiatric disorders.

Goldstein and Bukstein state that adolescence is a privileged stage for conducting comorbidity studies [40]. One important aspect would be the greater feasibility in determining what comes first, the MDs or substance abuse, and maybe what type of neurobiological changes could be the cause and what type could result in these disorders [40].

Again, based on the data available from studies in adult populations, one can assume with good chances of success that adolescents who already have MDs and become substance users or dependent will have greater difficulties in living with two comorbid diseases, which can affect each other, increase the level of morbidity, and possibly render treatments less effective. Additionally, individuals who have not yet developed the disease but are biologically vulnerable, in theory, could have the onset of symptoms that may not otherwise occur or could be delayed without substance abuse. Additionally, the influence of MDs on the damage caused by substance abuse could be deleterious, increasing morbidity and possible recovery. However, it is worth remembering that these possibilities need to be demonstrated through well-designed clinical studies conducted with samples of adolescents.

Information obtained from previous studies conducted in young populations support the inferences mentioned above, even without informative evidence.

The consumption of alcohol and other drugs among adolescents diagnosed with bipolar disorder appears to increase risks for suicide, unwanted pregnancies, abortions, and legal problems [41]. The comorbidity of depression and substance abuse seems to increase the suicide rates among adolescents [46].

Adolescents with comorbidity of bipolar disorder and substance abuse or dependence were less likely to continue living with their parents and presented a higher prevalence of posttraumatic stress disorder, conduct disorders, and obesity than adolescents who had bipolar disorder but had no problems with the use alcohol or other drugs[41, 47].

Data associating diseases with other events must be carefully used because there is not necessarily a causal relationship between occurrences of conditions detected together. The determination of the etiologic inter-relationship involves a demand for laborious evidence that is difficult to meet.

In a retrospective cohort study with 862 adolescents, Jerrel et al. showed that substance abuse preceded the onset of bipolar disorder in adolescence. However, this study did not detect significant differences in the course of illness associated with substance abuse or other comorbidities assessed in the study [48]. It is worth remembering that retrospective observational studies are more prone to bias and should be analyzed with restrictions.

Treatment of Comorbidity Between Mood Disorders and Substance Abuse or Dependence

The scientific literature still does not provide sufficient information to support interventions more appropriate to treat the comorbidity of MDs and substance abuse in adolescents. The studies demonstrating these interventions for adults are also very limited. Because these situations are very complex, multidisciplinary approaches involving psychiatrists, psychologists and, if possible, social service professionals should always be considered. The weight of each risk factor that contributes to the occurrence of substance abuse can vary, and this factor is important for planning the treatment. An existing MD needs to be adequately treated, which essentially means managed through pharmacological agents. Psychosocial interventions should be included in treatment, mainly the experimentally studied psychotherapies.

As already mentioned, the pharmacological treatment for children and adolescents is not identical to the treatment used for MDs in adults, either for cases of major depressive disorder or the bipolar spectrum of mood disorders. Although there are studies of pharmacological treatment for young people, they remain insufficient to end the controversy in this area [49]. Despite several open questions, especially about safety, antidepressants are used to treat depression in adolescents [50]. Lithium is still the only drug approved in the US to treat adolescents with bipolar disorder, but sodium valproate and carbamazepine are also used [51]. In the absence of sufficient data to justify treatments, the monitoring should be very careful.

Clinical studies on the effectiveness of specific pharmacological treatments for individuals with MDs who are also substance users are scarce for adults and even more so for adolescents. In adults, the effectiveness of lithium seems to decrease in cases of abuse of alcohol or other drugs [52]. A randomized placebo-controlled trial evaluated the use of sodium valproate in bipolar alcoholic patients, and the results suggest that this drug may be effective for both stabilizing mood symptoms and reducing alcohol use; the results showed a reduction in days of alcohol consumption, reduction of quantities consumed and longer time without relapsing[53]. Valproate has also been suggested as the most appropriate medication for this comorbidity because it treats bipolar disorder, reduces withdrawal symptoms, helps prevent relapse, does not present chemical properties that induce addiction mechanisms, and has good tolerability [54]. It is important that sodium valproate continue to be studied for treating this comorbidity, and it should also be extended to younger populations, such as adolescents.

A potential relevant problem is the difficulty of treatment adherence that seems to be common in this comorbidity. Bipolar patients have higher rates of inadequate treatment or non-adherence to mood-stabilizers [55].

Conclusion

Comorbidity between MDs and abuse of or dependence on substances seems to be high and involve greater difficulties for the control of each of these pathological conditions. The attention of healthcare professionals to this fact is crucial to minimize damage. Adolescents, although less studied than adults, are potentially more susceptible to both risk behaviors and more deleterious consequences in the case of comorbidity because they would have greater difficulties in acquiring the necessary adaptive resources. However, additional studies able to provide good quality evidence are required for more accurate discussions on this topic.

References

1. Associação Psiquiátrica Americana (APA). Manual Estatístico e Diagnóstico de Transtornos Mentais (DSM-IV-TR). 4th ed. Porto Alegre: Artmed; 2002.
2. Leboyer M, Henry C, Paillere-Martinot ML, et al. Age at onset in bipolar affective disorders: a review. Bipolar Disord. 2005;7(2):111–8.
3. Rao U, Chen L. Characteristics, correlates and outcomes of childhood and adolescent depressive disorder. Dialogues Clin Neurosci. 2009;11(1):45–62.
4. Davidson JR. Major depressive disorder treatment guidelines in America and Europe. J Clin Psychiatry. 2010;71(Suppl E1):e04.
5. Spijker J, Nolen WA. An algorithm for the pharmacological treatment of depression. Acta Psychiatr Scand. 2010;121(3):180–9.
6. Altamura AC, Lietti L, Dobrea C, et al. Mood stabilizers for patients with bipolar disorder: the state of art. Expert Rev Neurother. 2011;11(1):85–99.
7. Chang KD. Course and impact of bipolar disorder in young patients. J Clin Psychiatry. 2010;71(2):e-05.
8. Rodgers MJ, Zylstra RG, McKay JB, et al. Adolescent bipolar disorder: a clinical vignette. Prim Care Comp J Clin Psychiatry. 2010;12(4):PCC.09r00895.
9. Bechdolf A, Nelson B, Cotton SM, et al. A preliminary evaluation of the validity of at-risk criteria for bipolar disorders in help-seeking adolescents and young adults. J Affect Disord. 2010;127(1–3):316–20.
10. Nobile M, Cataldo GM, Marino C, et al. Diagnosis and treatment of dysthymia in children and adolescents. CNS Drugs. 2003;17(13):927–46.
11. Moreno C, Laje G, Blanco C, et al. National trends in the outpatient diagnosis and treatment of bipolar disorder in youth. Arch Gen Psychiatry. 2007;64(9):1032–9.
12. Bettge S, Wille N, Barkmann C, et al. BELLA study group. Depressive symptoms of children and adolescents in a German representative sample: results of the BELLA study. Eur Child Adolesc Psychiatry. 2008;17(Suppl 1):71–81.
13. Williamson DE, Birmaher B, Brent DA, et al. Atypical symptoms of depression in a sample of depressed child and adolescent outpatients. J Am Acad Child Adolesc Psychiatry. 2000;39(10):1253–9.
14. Patton GC, Coffey C, Posterino M, et al. Adolescent depressive disorder: a population-based study of ICD-10 symptoms. Aust N Z J Psychiatry. 2000;34(5):741–7.
15. Yorbik O, Birmaher B, Axelson D, et al. Clinical characteristics of syndromes in children and adolescents with major depressive disorder. J Clin Psychiatry. 2004;65(12):1654–9.
16. Gudienė D, Leskauskas D, Markeviciûté A, et al. Distinctions of bipolar disorder symptoms in adolescence. Medicina (Kaunas). 2008;44(7):548–52.
17. Wozniak J, Biederman J, Kiely K, et al. Mania-like symptoms suggestive of childhood-onset bipolar disorder in clinically referred children. J Am Acad Child Adolesc Psychiatry. 1995;34(7):867–76.
18. Joshi G, Wilens T. Comorbidity in pediatric bipolar disorder. Child Adolesc Psychiatr Clin N Am. 2009;18(2):291–319.
19. Birmaher B, Kennah A, Brent D, et al. Is bipolar disorder specifically associated with panic disorder in youths? J Clin Psychiatry. 2002;63(5):414–19.
20. Kim EY, Miklowitz DJ. Childhood mania, attention deficit hyperactivity disorder and conduct disorder: a critical review of diagnostic dilemmas. Bipolar Disord. 2002;4(4):215–25.
21. Geller B, Tillman R. Prebubertal and early adolescent bipolar I disorder: review of diagnostic validation by Robins and Guze. J Clin Psychiatry. 2005;66 (Suppl 7):21–8.
22. Perlis RH, Miyahara S, Marangell LB, et al. STEP-BD Investigators. Long-term implications of early onset in bipolar disorder: data from the first 1000 participants in the systematic treatment enhancement program for bipolar disorder (STEP-BD). Biol Psychiatry. 2004;55(9):875–81.

23. Regier DA, Farmer MF, Rea DS, et al. Comorbidity of mental disorders and alcohol or drug abuse. Results of Epidemiologic Catchement Area (ECA) Study. JAMA. 1990;234(19):2511–18.
24. Compton WM, Thomas YF, Stinson FS, et al. Prevalence, correlates, disability, and comorbidity of DSM-IV drug abuse and dependence in the United States: results from the national epidemiologic survey on alcohol and related conditions. Arch Gen Psychiatry. 2007;64(5):566–76.
25. Conway KP, Compton W, Stinson FS, et al. Lifetime comorbidity of DSM-IV mood and anxiety disorders and specific drug use disorders: results from the national epidemiologic survey on alcohol and related conditions. J Clin Psychiatry. 2006;67(2):247–57.
26. Rao U, Ryan ND, Dahl RE, et al. Factors associated with the development of substance use disorder in depressed adolescents. J Am Acad Child Adolesc Psychiatry. 1999;38(9):1109–17.
27. Wilens TE, Biederman J, Kwon A, et al. Risk of substance use disorders in adolescents with bipolar disorder. J Am Acad Child Adolesc Psychiatry. 2004;43(11):1380–6.
28. Ernst M, Fudge J. A developmental neurobiological model of motivated behavior: anatomy, connectivity and ontogeny of triadic nodes. Neurosci Biobehav Res. 2009;33(3):367–82.
29. Holden B, Gitlesen JP. The relation between psychiatric symptomatology and motivation of challenging behavior: a preliminary study. Res Dev Disabil. 2007;29(5):408–13.
30. Swann AC. The strong relationship between bipolar and substance-use disorder. Ann N Y Acad Sci. 2010;1187:276–93.
31. McElroy SL, Pope HG Jr, Keck PE Jr, et al. Are impulse-control disorders related to bipolar disorder? Compr Psychiatry. 1996;37(4):229–40.
32. Bizzarri JV, Rucci P, Sbrana A, et al. Reasons for substance use and vulnerability factors in patients with substance use disorder and anxiety or mood disorders. Addict Behav. 2007;32(2):384–91.
33. Brady KT, Sinha R. Co-occurring mental and substance use disorders: the neurologic effects of chronic stress. Am J Psychiatry. 2005;162:1483–93.
34. Tremblay LK, Naranjo CA, Cardenas L, et al. Probing brain reward system function in major depressive disorder: altered response to dextroamphetamine. Arch Gen Psychiatry. 2002;59:409–16.
35. Schlaepfer TE, Cohen MX, Frick C, et al. Deep brain stimulation to reward circuitry alleviates anhedonia in refractory major depression. Neuropsychopharmacology. 2008;33(2):368–77.
36. Wacker J, Dillon DG, Pizzagalli DA. The role of the nucleus accumbens and rostral anterior cingulate cortex in anhedonia: integration of resting EEG, fMRI, and volumetric techniques. Neuroimage. 2009;46(1):327–37.
37. Johnson C, Drgon T, McMahon FJ, et al. Convergent genome wide association results for bipolar disorder and substance dependence. Am J Med Genet B Neuropsychiatr Genet. 2009;150B(2):182–90.
38. Sajatovic M, Levin J, Fuentes-Casiano E, et al. Illness experience and reasons for non-adherence among individuals with bipolar disorder who are poorly adherent with medication. Compr Psychiatry. 2011;52(3):280–7.
39. Cassidy F, Ahearn EP, Carroll BJ. Substance abuse in bipolar disorder. Bipolar Disord. 2001;3(4):181–8.
40. Goldstein BI, Bukstein OG. Comorbid substance use disorders among youth with bipolar disorder: opportunities for early identification and prevention. J Clin Psychiatry. 2010;71(3):348–58.
41. Goldstein BI, Strober MA, Birmaher B, et al. Substance use disorders among adolescents with bipolar spectrum disorders. Bipolar Disord. 2008;10(4):469–78.
42. Geller B, Tillman R, Bolhofner K, et al. Child bipolar I disorder: prospective continuity with adult bipolar I disorder; characteristics of second and third episodes; predictors of 8-year outcome. Arch Gen Psychiatry. 2008; 65(10):1125–33.
43. Wittchen HU, Fröhlich C, Behrendt S, et al. Cannabis use and cannabis use disorders and their relationship to mental disorders: a 10-year prospective-longitudinal community study in adolescents. Drug Alcohol Depend. 2007;88(Suppl 1):60–70.

44. O'Neil KA, Conner BT, Kendall PC. Internalizing disorders and substance use disorders in youth: comorbidity, risk, temporal order, and implications for intervention. Clin Psychol Ver. 2011;31(1):104–12.
45. Goldberg JF, Garno JL, Callahan AM, et al. Overdiagnosis of bipolar disorder among substance use disorder inpatients with mood instability. J Clin Psychiatry. 2008;69(11): 1751–7.
46. Deas D. Adolescent substance abuse and psychiatric comorbities. J Clin Psychiatry. 2006;67(Suppl 7):18–23.
47. Goldstein BJ, Birmaher B, Axelson DA, et al. Preliminary findings regarding overweight and obesity in pediatric bipolar disorder. J Clin Psychiatry. 2008;69(12):1953–9.
48. Jerrell JM, McIntyre RS, Tripathi A. A cohort study of the prevalence and impact of comorbid medical conditions in pediatric bipolar disorder. J Clin Psychiatry. 2010;71(11):1518–25.
49. Thomas T, Stansifer L, Findling RL. Psychopharmacology of pediatric bipolar disorders in children and adolescents. Pediatr Clin North Am. 2011;58(1):173–87.
50. Smiga SM, Elliott GR. Pharmacology of depression in children and adolescents. Pediatr Clin North Am. 2011;58(1):175–81.
51. Madaan V, Chang KD. Pharmacotherapeutic strategies for pediatric bipolar disorder. Expert Opin Pharmacother. 2007;8(12):1801–19.
52. Goldberg JF, Garno JL, Leon AC, et al. A history of substance abuse complicates remission from acute mania in bipolar disorder. J Clin Psychiatry. 1999;60(11):733–40.
53. Salloum IM, Cornelius JR, Daley DC, et al. Efficacy of valproate maintenance in patients with bipolar disorder and alcoholism: a double-blind placebo-controlled study. Arch Gen Psychiatry. 2005;62(1):37–45.
54. Kosten TR, Kosten TA. New medication strategies for comorbid substance use and bipolar affective disorders. Biol Psychiatry. 2004;56(10):771–7.
55. Manwani SG, Szilagyi KA, Zablotsky B, et al. Adherence to pharmacotherapy in bipolar disorder patients with and without co-occurring substance use disorders. J Clin Psychiatry. 2007;68(8):1172–6.

Luís Pereira Justo MS in Psychobiology, Federal University of São Paulo (UNIFESP). MD, Federal University of Rio de Janeiro (*Universidade Federal do Rio de Janeiro*—UFRJ). Researcher, Associação Fundo de Incentivo à Pesquisa.

Helena Maria Calil MD School of Medical Sciences of the Holy House of Mercy of São Paulo (*Faculdade de Ciências Médicas da Santa Casa de São Paulo*—FCMSC-SP). PhD in Pharmacology, specializing in Psychopharmacology, Federal University of São Paulo/School of Medicine of São Paulo (UNIFESP/EPM); Postdoctoral Research Fellow at Stanford University, USA. Visiting Fellow and Visiting Scientist at the National Institute of Mental Health, USA. Habilitation in Clinical Psychiatry, Department of Psychiatry, Discipline of Clinical Psychiatry, UNIFESP/EPM. Full Professor, Psychopharmacology, Department of Psychobiology, UNIFESP/EPM. Director, Center of Research in Clinical Psychobiology. Founding Partner, Brazilian Psychobiology Society (currently, Neuroscience and Behavior Society), Latin American College of Neuropsychopharmacology (current member of the Advisory Council), and Brazilian Bipolar Disorder Association (current member of the Supervisory Board). Member of the Society of Biological Psychiatry, the Society of Light Therapy and Biological Rhythms, the Collegium Internationale Neuro-Psychopharmacologicum, and the Brazilian Psychiatric Association. Vice-President, Psychopharmacology Incentive Fund Association.

Chapter 12
Patterns of Drug Use in Carriers of Attention Deficit and Hyperactivity Disorder (ADHD)

Alfredo Löhr and Marco Antonio Bessa

Background

Attention deficit hyperactivity disorder (ADHD) is a neurobiological syndrome that begins to manifest at an early age. It extends into adolescence and reaches adulthood, although the clinical manifestations change over time. According to the American Academy of Pediatrics, this neurobiological disorder is more common among children. In the school–age population, its estimated frequency is between 5 % and 12 % [1]. Recent studies suggest that ADHD can persist into adulthood at a frequency of 10 to 60 % [2–4]. Children with this disorder exhibit early symptoms of inappropriate hyperactivity, inattention, low academic performance, and impulsive behavior as well as increased risk of delinquency, substance abuse, and accidents. Disruptive behaviors are the main reasons for seeking treatment, but associations with other psychopathologies are frequent, such as conduct disorder, mood disorders, and substance use disorders (SUD). There is considerable knowledge that supports the idea that ADHD is a familial disorder associated with differences in the structure, metabolism, and processing of the central nervous system [4]. Several studies have directly correlated ADHD with substance abuse during childhood, adolescence, and throughout adult life [5, 6]. Children with ADHD monitored during the transition to adolescence exhibited higher use of alcohol, tobacco, and other drugs compared with non-carriers of ADHD [7].

However, Bukstein [8] highlights that despite the importance of the association between ADHD and SUD and the recommendation to screen of adolescents for

A. Löhr (✉)
Center of Health Sciences, Pontifical Catholic University of Paraná (Pontifícia Universidade Católica do Paraná—PUCPR), Curitiba, Brazil
e-mail: alfredo.lohr@pucpr.br

M. A. Bessa
Federal University of São Paulo/School of Medicine of São Paulo (UNIFESP/EPM), São Paulo, Brazil
e-mail: marco.bessa@uol.com.br

D. De Micheli et al. (eds.), *Drug Abuse in Adolescence*,
DOI 10.1007/978-3-319-17795-3_12

alcohol and drug abuse during primary health care, the American Academy of Pediatrics estimates that less than 50 % of pediatricians perform this procedure.

ADHD Versus Drug Abuse

There is a strong association between ADHD and drug abuse. First, it has been known that dopamine (DP) is an important neurotransmitter in the central nervous system and that it participates in the neurotransmission mechanisms related to ADHD [9]. Previous studies using animal models [10] suggest a deficit in the inhibitory response and indicate a dysfunction in the ventrolateral frontal cortex, cingulate, and basal ganglia in ADHD and substance abuse. Recent neuroimaging studies using positron emission tomography (PET) in subjects with ADHD with a history of drug abuse showed decreased DP levels in the striatal region and a dysfunction between the cingulate cortex and the prefrontal cortex. Animal studies conducted with electrophysiological techniques to assess the electrical activity of dopaminergic neurons and microdialysis to measure DP in mesolimbocortical circuits have found a common pattern of activity of drugs of abuse. These substances can increase the bioavailability of synaptic DP in mesolimbocortical nerve endings [11]. Psychostimulant substances increase the effect of DP by inhibiting its reuptake. Alcohol acts on the ventral tegmental area, indirectly increasing the frequency of discharges of dopaminergic neurons by attenuating the inhibitory tone on GABAergic interneurons. Opioid substances, through [micron]–opioid receptors located in the somatodendritic area of the ventral tegmental region, also increase dopaminergic discharges. Similarly, nicotine causes increased neurophysiological activity of dopaminergic neurons in the ventral tegmental area.

From the genetic perspective, ADHD and drug abuse share many factors in common. The relationship among these factors has its origin in the brain dopaminergic system, which, through genetic mechanisms, affects the susceptibility of the DP–releasing circuits. Allelic variants of the catechol-O-methyltransferase (COMT) locus are predisposed to genetic differences that, to a greater or lesser extent, determine susceptibility to drug abuse [12]. A decrease in the prefrontal lobe volume has also been observed in drug users, although it cannot be concluded whether this anatomical dysfunction is primary or secondary to drug abuse [13]. Other hypotheses have been proposed to try to explain the increased risk of substance abuse among ADHD patients. Personality traits with genetic mediation, such as impulsivity and novelty-seeking behavior, are characteristics common to both ADHD and SUD and may have the same neurological basis and thereby help establish an association [14]. Furthermore, it has been proposed that carriers of ADHD consume addictive drugs as self-medication for their symptoms and that the impulsivity associated with this disorder contributes to the development of dependence [15].

Moreover, children or adolescents with ADHD, particularly the hyperactive–impulsive and mixed subtypes, demonstrate significant academic and social dysfunction, characterized by impulsivity and limited inhibitory control. They participate in fights and are intolerant of frustration, experience poor academic performance, and have high dropout rates. As a result, these children tend to suffer criticism and punishment from adults and to be stigmatized and discriminated against in their social environment. This condition can lead adolescents to seek support from others living in similar conditions or among drug users. The drug, which at first may satisfy a simple curiosity about new sensations for individuals who need stimulation and new experiences, can later bring symptom relief and a sense of well-being and, with repeated use, can lead to addiction with all its consequences [16].

Approximately 40 to 60 % of children with ADHD retain symptoms into adolescence and adulthood and present a high probability of developing comorbid conditions such as SUD, e.g., alcohol abuse (32 to 53 %) and marijuana and cocaine use (8 % and 32 %, respectively) [17].

Several studies that evaluated patients with SUD and ADHD showed that they initiate drug use earlier, have a higher probability of increased consumption and multiple substance abuse, and develop dependence more often. Compared to drug users without ADHD, these patients have increased hospitalization rates for drug abuse [18]. In addition, it was found that adolescents with ADHD have a higher risk of alcohol, tobacco, and illicit drug use compared to adolescents without ADHD. Patients with this comorbidity show a lower adherence to SUD treatment programs, lower rates of remission from substance abuse, and a more prolonged course of drug use [19].

A study involving adults diagnosed with ADHD in childhood found that approximately 40 % of these patients abused or showed dependence on psychoactive substances during their lifetime, and regardless of comorbidities with other psychiatric disorders, the risk of substance abuse among patients with ADHD was higher [20].

Longitudinal studies indicate that the onset of ADHD precedes substance use. Therefore, it seems that ADHD in most patients is not secondary to drug use [15].

ADHD and Drug Abuse

ADHD and Nicotine Use

Children with ADHD are twice as likely to become nicotine smokers than non–carriers of ADHD. In adolescence, this probability reaches threefold [21]. Adolescents with ADHD with a smoking history have an increased risk of posterior abuse of alcohol and other drugs compared to patients with ADHD without a smoking history [22]. A 10-year follow-up found that ADHD is a significant predictor of any SUD, including tobacco use. The results also indicate that ADHD is a significant risk factor for the development of SUD and cigarette smoking in both sexes [23].

ADHD and Alcohol Abuse

Some studies have suggested that low levels of DP or the high density of free dopamine D2 receptors in the striatum could be related to the early release of alcohol-dependent patients [24]. Another study shows that 14% of adolescents between 15 and 17 years with ADHD have problems with alcohol abuse or dependence compared to their non-carrier counterparts. Specifically, children with ADHD have 1.7 times more probability of engaging in alcohol abuse [25]. A Turkish study by Ercan et al. [26] showed that adolescents with ADHD begin alcohol dependence earlier. In addition, ADHD in children is associated with alcohol and drug abuse into adulthood and the use of nicotine in adolescence [27, 28].

ADHD and Marijuana Use

Previous studies have found a strong correlation between the use of marijuana and other "recreational" drugs and psychological disorders, such as obsessive–compulsive disorder. Children with ADHD have 1.5 times more probability of using and becoming dependent on marijuana compared to young non-carriers of ADHD [25].

ADHD and Cocaine Use

Wilens et al. [28] found that the rate of children whose parents were cocaine addicts with an ADHD score in the Child Behavior Checklist (CBCL) subscale was 23%. A strong correlation was observed between cocaine use in adulthood and ADHD. Previous studies involving adult drug users (usually cocaine users) found an ADHD prevalence of approximately 35%, which was significantly higher than the expected rate in the general population [29, 30]. A previous study indicated that children with ADHD are twice as likely to develop cocaine dependence than non-carriers of ADHD [25]. Among patients subjected to treatment for cocaine use, the prevalence of ADHD in childhood was 5%–35% [18].

Other studies have shown that 35% of cocaine addicts have a history of ADHD, and approximately 15% of cocaine users may suffer from ADHD in their adult life. In turn, children with ADHD who received treatment with psychostimulants have three to four times lower likelihood of abusing drugs than the untreated group [9]. High consumption of drugs has also been observed in the relatives of children with ADHD.

ADHD, Conduct Disorder, and Drug Use

In adolescents, predominantly in male subjects with conduct disorders, ADHD and depression are important comorbidities related to substance dependence [31, 32]. It is known that the comorbidity between ADHD and conduct disorder is frequent and that the latter, in turn, is clearly associated with drug abuse and addiction. In this respect, SUD may occur more frequently in the subgroup of adolescents with ADHD who jointly have conduct disorder. Therefore, the comorbidity involving conduct disorder would be the risk factor, rather than ADHD itself. However, this issue needs to be further elucidated [33]. Another study shows that hyperactivity/impulsivity can predict later problems related to substance use even when considering late-onset conduct disorders. Inattention alone has a lower risk of developing ADHD, and a single ADHD symptom is associated with higher risk [34].

Therapeutic Approach

A review of the association between ADHD and SUD in children and adolescents and the treatment of this comorbidity suggests that drug therapy for ADHD does not increase the risk for subsequent SUD, although psychostimulants may be abused by patients and families. Other non-stimulant psychotropic medications, including atomoxetine, have low abuse potential but seem to be less effective than stimulants. Novel treatments with drugs having decreased potential for abuse are being developed, e.g., a transdermal form of methylphenidate and lisdexamfetamine dimesylate (LDX), a stimulant pro-drug recently approved for the treatment of ADHD. The authors conclude that drug therapy can decrease the risk of SUD among carriers of ADHD, and stimulants remain the first–line therapy for treatment of major ADHD symptoms [35].

A meta–analysis of six studies involving children with ADHD treated with stimulants concluded that SUD was less frequent in adolescence in the treated group compared to children with ADHD not subjected to drug therapy. However, the protective action of the drugs appears to continue through adulthood. Therefore, the best strategy to decrease the development of SUD is to provide effective and early treatment of ADHD [36].

References

1. American Academy of Pediatrics. Clinical practice guideline: diagnosis and evaluation of the child with attention-deficit/hyperactivity disorder. Pediatrics. 2010;105:1158–70.
2. Scahill L, Schwab-Stone M. Epidemiology of ADHD in school-age children. Child Adol Psychiatr Clin N Am. 2000;9:541–55.
3. Swanson JM, Sergeant JA, Taylor E, et al. Attention-deficit hyperactivity disorder and hyperkinetic disorder. Lancet. 1998;351:429–33.
4. Polanczyk G, Lima MS, Horta BL, et al. The worldwide prevalence of ADH D: a systematic review and metaregression analysIs. Am Psychiatry. 2007;164:942–8.
5. Biederman J, Wilens TE, Mick E, et al. Does attention-deficit hyperactivity disorder impact the development of drug and alcohol abuse and dependence? Biol Psychiatry. 1998;44(4):269–7.

6. Willens TE. The nature of relationship between attention deficit/hyperactivity disorder and substance use. J Clin Psychiatry. 2007;68(Suppl. 11):4–8.
7. Biederman J, Molina BS, Pelham WE. Childhood predictors of adolescent substance abuse in a longitudinal study of children with ADHD. J Abnorm Psychol. 2003;112:497–507.
8. Bukstein O. Substance abuse in patients with attention-deficit/hyperactivity disorder. Medscape J Med. 2008;10(1):24.
9. Levin FR, Evans SM, McDowell DM, et al. Methylphenidate treatment cocaine abusers with adult attention deficit/hyperactivity disorder: a pilot study. J Clin Psychiatry. 1998;59:300–5.
10. Vendruscolo LF, Takahashi RN. Comorbidade entre o transtorno de déficit de atenção/hiperatividade e o abuso e dependência de álcool e outras drogas: evidências por meio de modelos animais. Rev Bras Psiquiatr. 2011;33(2):203–8.
11. Cubillo A, Halari R, Smith A, et al. A review of frontostriatal and fronto-cortical brain abnormalities in children and adults with Attention Deficit Hyperactivity Disorder (ADHD) and new evidence for dysfunction in adults with ADHD durimotivation and attention. Cortex. 2011;30:1–22.
12. Van den Bergh DJ, Rodriguez LA, Miller IT, et al. High-activity cathechol-O-methyl-transferase allele is more prevalent in polysubstance abusers. Am J Med Genet. 1997;74:439–42.
13. Looby A. Childhood attention deficit hyperactivity disorder and the development of substance use disorders: valid concern or exaggeration? Add Behav. 2008;33:451–63.
14. Chambers RA, Taylor JR, Potenza MN. Developmental neurocircuitry of motivation in adolescence: a critical period of addiction vulnerability. Am J Psychiatry. 2003;160:1041–52.
15. Wilens TE, Biederman J. Alcohol, drugs, and attention-deficit/hyperactivity disorder: a model for the study of addictions in youth. J Psychopharmacol. 2006;20:580–8.
16. Bessa MA, Boaratti MA, Scivoletto S. Crianças e adolescentes. In: Diehl A, Cordeiro DC, Laranjeira R, Editors. Dependência química: prevenção, tratamento e políticas públicas. Porto Alegre: Artmed; 2010, p. 359–74.
17. Wilens TE, Adamson J, Monuteaux MC, et al. Effect of prior stimulant treatment for attention-deficit/hyperactivity disorder on subsequent risk for cigarette smoking and alcohol and drug use disorders in adolescents. Arch Pediatr Adolesc Med. 2008;162(10):916–2.
18. Scivoleto S. Drogas e álcool. In: Assumpção FB, Kaczynski E, Editors. Tratado de psiquiatria da infância e adolescência. São Paulo: Atheneu; 2003, p. 525.
19. Wilens TE, Upadhyaya HP. Impact of substance use disorder on ADHD and its treatment. J Clin Psychiatry. 2007;68(8):20.
20. Biederman J, Wilens T, Mick E, et al. Psychoactive substance use disorders in adults with attention deficit hyperactivity disorder (ADHD): effects of ADHD and psychiatric comorbidity. Am J Psychiatry. 1995;152:1652–8.
21. Wilens TE, Vitulano M, Upadhyaya H, et al. Important environmental factors appear to add to the risk of smoking associated with ADHD. J Pediatr. 2008;153:414–9.
22. Biederman J, Monuteaux MC, Mick E, et al. Is cigarette smoking a gateway to alcohol and illicit drug disorders? A study of youths with and without attention deficit hyperactivity disorder. Biol Psychiatry. 2006;59:258–64.
23. Wilens TE, Martelon M, Joshi G, et al. Does ADHD predict substance-use disorders? A 10-year follow-up study of young adults with ADHD. J Am Acad Child Adolesc Psychiatry. 2011;50(6):543–53.
24. Guardia J, Catafau AM, Beatle F, et al. Striatal dopaminergic D2 receptor density measured by {I123} Iodobenzamide SPECT in the prediction of treatment outcome of alcohol-dependents patients. A J Psychiatry. 2000;157:127–9.
25. Lee SS, Humphreys KL, Flory K, et al. Prospective association of childhood attention-deficit/hyperactivity disorder (ADHD) and substance use and abuse/dependence: a meta-analytic review. Clin Psychol Rev. 2011;31:328–41.
26. Ercan ES, Coskunol H, Varan A, et al. Childhood attention deficit/hyperactivity disorder and alcohol dependence: a 1-year follow-up. Alcohol & Alcoholism. 2003;38(4):352–6.

27. Charac A, Yeung E, Climans T, et al. Childhood attention-deficit/hyperactivity disorder and future substance use disorders: comparative meta-analyses. J Am Acad Child Adolesc Psychiatry. 2011;50(1):9–21.
28. Wilens T, Biederman J, Kiely K, et al. Pilot study of behavior and emotional disturbances in high-risk children of parents with opioid dependence. J Am Acad Child Adol Psychiatry. 1995;34:779–85.
29. Brook D, Brook JS, Zhang C, et al. The association between attention deficit hyperactivity disorder in adolescence and substance use disorders in adulthood. Arch Pediatr Adolesc Med. 2010;164(10):930–4.
30. Vergara-Moraguesa E, González-Saiza F, Lozano-Rojasc O, et al. Diagnóstico del trastorno por déficit de atención e hiperactividad (TDAH) en pacientes adultos com dependencia de cocaína: utilidad de los nuevos síntomas de funcionamiento ejecutivo de Barkley. Drug Alcohol Depend. 1997;47:87–97.
31. Glass K, Flory K, Martin A, et al. ADHD and comorbid conduct problems among adolescents: associations with self-esteem and substance use. ADHD Atten Def Hyp Disord. 2011;3:29–39.
32. Molina BSG, Smith BH, Pelham W. Interactive effects of attention deficit hyperactivity disorder and conduct disorder on early adolescent substance use. Psychol Add Behav. 1999;13:348–58.
33. Rohde LAM, Barbosa G, Tramontina S, et al. Transtorno de déficit de atenção/hiperatividade. Rev Bras Psiquiatr. 2000;22(suppl. 2):7–11.
34. Elkins IJ, McGue M, Iacono WG. Prospective effects of attention-deficit/hyperactivity disorder, conduct disorder, and sex on adolescent substance use and abuse. Arch Gen Psychiatry. 2007;64(10):1145–52.
35. Upadhyaya HP. Substance use disorders in children and adolescents with attention-deficit/hyperactivity disorder: implications for treatment and the role of the primary care physician. Prim Care Companion J Clin Psychiatry. 2008;10(3):211–21.
36. Wilens TE, Faraone SV, Biederman J, et al. Does stimulant therapy of attention-deficit/hyperactivity disorder beget later substance abuse? A meta-analytic review of the literature. Pediatrics. 2003;111:179–85.

Alfredo Löhr Junior Associate Professor of Medicine, Center of Health Sciences, Pontifical Catholic University of Paraná (Pontifícia Universidade Católica do Paraná—PUCPR). MA, Internal Medicine, Department of Neurology, Paraná Federal University Teaching Hospital (Universidade Federal do Paraná—UFPR). Chief of Pediatric Neurology, Pequeno Príncipe Hospital, Curitiba.1,*,

Marco Antonio Bessa Psychiatrist. Specialist in Chemical Dependency, Federal University of São Paulo/School of Medicine of São Paulo (UNIFESP/EPM). Doctor of Science, Department of Psychiatry, UNIFESP/EPM. MA in Philosophy, Federal University of São Carlos (Universidade Federal de São Carlos—UFSCar). Researcher, National Institute of Science and Technology for Policies on Alcohol and Other Drugs. Councilor, Regional Medical Council of Paraná.

Index

D. de Micheli et al. (eds.), *Drug Abuse in Adolescence*,
DOI 10.1007/978-3-319-17795-3

MIX
Papier aus verantwortungsvollen Quellen
Paper from responsible sources
FSC® C105338

Printed by Books on Demand, Germany